Design of Liquid-Fueled Rocket Engines

Volume 1

Dr. James R., Ph.D.

Table of Contents

Volume 5

Chapter 1

In this book, Tutor in Book's Design of Liquid-Fueled Rocket Engines, the student will learn about the various types of rocket missiles and their engines. In a missile, the basic element is always the engine.

The type of engine is determined basically by the design, overall dimensions, and flight dimensions of these missiles.

The Missile's range depends on the type of engine selected, the type of fuel components used in it, and its design, operational, and other characteristics.

It is the object of the present volume to provide students in the higher technical educational institutions and engineering-technical workers with an outline of the theory and basic design principles of liquid-fueled rocket engines.

as well as to know the chemistry and combustion theory of liquid fuels, the fundamentals of heat exchange and heat transfer, the theory of strength and rigidity of structures, high-speed aerodynamics, ballistics, and automatic-control theory.

SECTION 1. BRIEF HISTORY OF THE DEVELOPMENT OF LIQUID-FUELED ROCKET ENGINES

Our country can take the basic credit for developing the foundations of the theory of liquid-fueled rocket engines and the missiles for them, as well as for building the first models.

The founder of reaction-engine theory was the great Russian scientist N.Ye. Zhukovskiy, who, in his 1882 paper entitled "O reaktsii vytekayushchey i vtekayushchey zhidkosti" ("Concerning the reaction of outflowing and inflowing liquids"), was the first to derive the equation for the reaction force of an outflowing fluid, and who presented the equation for the efficiency of the fluid stream in his later papers.

The Russian scientist I.V. Meshcherskiy, whose outstanding works on the mechanics of bodies of variable mass, which were published in 1897, formed the basis for the contemporary theory of motion of reaction-thrust missiles, must be ascribed an important part in the creation of reaction-engine theory.

K.E. Tsiolkovskiy, who originated the idea of the liquid-fueled rocket engine, created the theory of flight of reaction-powered machines, and authored a number of systems for such machines and other suggestions, must be included among the Russian scientists who have made the inventive genius of their Motherland famous.

K.E. Tsiolkovskiy published his first paper in the journal "Nauchnoye obozreniye" ("Scientific Review"), No. 5, under the title "Issledovaniye mirovykh prostranstv reaktivnymi priborami" ("Investigation of interstellar space by means of reaction-engined devices"). In this 1903 paper, and in the supplementary papers that followed it between 1911 and 1914, and then in 1924-1925 and thereafter, K. E. Tsiolkovskiy showed the insight of a genius in formulating and solving the basic problems of the theory and working principles of the liquid-fueled rocket missile.

He was first to propose the following:

1) a design for a liquid-fueled rocket engine (1903);

2) the use of liquid oxygen as an oxidizer for the fuel (1903);

3) pump delivery of the fuel into the engine's chamber (1903);

4) graphite and high-melting metals for fabrication of engine chambers (1903);

5) cooling of the engine chamber by one of the fuel components (1903);

6) closed-cycle cooling of the engine chamber (1903);

7) the use of nuclear and electric energy to power reaction engines (1911);

8) the use of ozone as an oxidizer and liquid methane and turpentine as fuels (1914);

9) the use of oxides of nitrogen as oxidizers (1926);

10) the use of nitrogen pentoxide as an oxidizer and monatomic hydrogen as a fuel (1927);

11) placing control surfaces in the stream of gases behind the nozzle of the engine chamber to control the missile in its flight in the rarefied layers of the atmosphere and in space (1929), etc.

K. E. Tsiolkovskiy also considered other problems of reaction-thrust motion in his papers, and specifically:

1) he demonstrated that a reaction-thrust engine can develop thrust even in airless space;

2) he made reference to the high promise of practical utilization of reaction-thrust missiles for study of cosmic space;

3) he demonstrated that an inclined ascent is more efficient for a reaction-thrust missile than vertical ascent;

4) he proved that liquid fuels are superior to solid fuels (powder) for medium- and long-range reaction-thrust machines;

5) he showed that medium- and long-range reaction-thrust missiles must have control and automatic-control devices in operation during flight;

6) he studied the motion of the reaction-thrust missile _in vacuo_, with and without consideration of gravity;

7) he derived a celebrated formula for determining the maximum flight speed of a reaction-thrust missile in airless space at the end of its powered trajectory;

8) he made note of the expediency of raising the gas pressure in the engine's combustion chamber with the object of increasing its specific thrust;

9) he determined the efficiency of the reaction-thrust engine during flight and indicated a number of paths for development of this type of technique and the promise that it offered.

K. E. Tsiolkovskiy established a solid scientific footing for the study of the rocket's motion as a body of variable mass.

K. E. Tsiolkovskiy advanced the following ingenious ideas:

1) that of building multiple-stage reaction-thrust missiles to attain long flight ranges;

2) the expediency of using winged reaction-thrust missiles to attain long ranges;

3) the design of reaction-engined aircraft;

4) the idea of cosmic stations, i.e., artificial Earth satellites

in the form of specialized reaction-thrust missiles, etc.

Contemporary reaction-thrust technology is based for the most part on the work of K.E. Tsiolkovskiy. His ideas were popularized to some extent in the book "Mezhplanetnyye puteshestviya" ("Interplanetary Travel") by Ya. I. Perel'man, which appeared in 1915.

K.E. Tsiolkovskiy began to show interest in problems of reaction-thrust motion in 1883. Between February and May of that year, he wrote a paper entitled "Svobodnoye prostranstvo" ("Free space"), in which he analyzed certain problems related to the use of reaction-thrust motion. In 1896, he wrote a novel entitled "Vne zemli" ("Outside the earth"), in which he also made reference to the reaction-thrust missile as an apparatus for interplanetary travel.

The Great October Socialist Revolution gave K.E. Tsiolkovskiy a full opportunity to develop his creative activity. The daily attention paid to his scientific research work by the Communist Party and the Soviet Government contributed to its widespread dissemination and recognition.

Before the Great October Socialist Revolution, he had written 80 papers, of which 50 had been published, while after the revolution, he wrote and published 150 articles.

The numerous works of K.E. Tsiolkovskiy in the field of reaction-engine technology made him world-famous. Professor H. Oberth wrote K.E. Tsiolkovskiy from Germany in September of 1929 as follows: "My compliments... May you live to see the attainment of your glorious goals... You have lighted the lamp, and we shall work until the most grandiose dream of humanity comes true."

In 1917, the Russian scientist Yu.M. Kondratyuk began his research work in the field of reaction-thrust motion, and in 1929 he published a paper entitled "Zavoyevaniye mezhplanetnykh prostranstv"

("Conquest of interplanetary space") (republished by Oborongiz in 1947). In this paper, he derived a series of important conclusions and gave expression to many original ideas in the field of reaction-thrust technique:

1) he demonstrated that a missile that does not jettison its fuel tanks during flight as they are exhausted cannot escape the range of terrestrial gravity;

2) he investigated the takeoff dynamics of the winged reaction-thrust missile;

3) he introduced new proposals on the realization of space stations and interplanetary flights using reaction-thrust machines;

4) he proposed the use of solids (lithium, boron, etc.) as fuels for ZhRD.

He also investigated the problem of heating of reaction-thrust missiles during flight in the atmosphere.

Our compatriot F.A. Tsander began his studies of reaction propulsion in 1908, and in his 1932 book entitled "Problemy poleta pri pomoshchi reaktivnykh apparatov" ("Problems of flight using reaction-thrust machines") set forth the results of his investigations of a number of highly important problems of reaction-thrust engineering and mapped out paths for its development, as follows:

1) he made note of the expediency of using metallic elements of the reaction-thrust missile as fuel after they had performed their immediate functions, with the object of attaining maximum range;

2) he developed a theory of spray injectors and a theory of the ZhRD, analyzing the basic principles of heat-exchange calculation and cooling for the engine chamber;

3) he proposed a number of operating cycles for the ZhRD and derived their efficiencies;

4) he developed a theory of missile flight along elliptical trajectories;

5) he investigated the climbing properties of a reaction-engined aircraft;

6) he investigated various types of liquid fuels with the object of selecting the most efficient fuels for use in reaction-thrust engines, and considered a number of other problems related to reaction-thrust engineering.

In 1930-1932, F.A. Tsander built and tested his OR-1 reaction engine, which worked on a mixture of air and gasoline and developed up to 5 kg of thrust. A second experimental engine, the OR-2, which had been built after his drawings in 1932 for operation on liquid oxygen and gasoline, underwent firing tests in 1933 without participation of F.A. Tsander, who had died ten days after the first test of the engine. The combustion chamber of this engine was oxygen-cooled, and the nozzle was water-cooled. The fuel components were fed into this engine's combustion chamber by compressed air.

Abroad, the first investigations of cosmic-flight problems appeared beginning with the second decade of the 20th century, and these were followed by studies of liquid-fueled rocket missiles. Among the scientists who devoted their efforts to these problems, we should take note of R. Éno-Peltri (France), whose first paper was published in 1903, and R. Goddard (USA), who began his work in 1915 and subsequently built several types of meteorological missiles with liquid-fueled engines. The work of H. Oberth (Germany), which was published in 1927, and that of E. Zenger (Austria), which appeared in 1933, made large contributions to the theory of reaction-engined flight.

The first dependable ZhRD were built in the USSR. The engines of the reaction-thrust institute at Peenemunde (the A-4 and "Wasserfall") and the firms BMW [Bayerische Motoren Werke] and Walther appeared in Germany, and those of the California Institute of Technology and the firms Aerojet Engineering, Reaction Motors, etc. in the USA all appeared somewhat later than the first ones built in the USSR.

The appearance of the first dependable engines made definite the possibility of applying them in practice in long-range and antiaircraft missiles of various types, aerial torpedos, etc.

A tremendous achievement in the field of developing reaction-thrust science and technology was the building and successful testing of the world's first intercontinental ballistic missile in the Soviet Union.

Test results for this rocket indicated that perfection of the intercontinental ballistic missiles will make it possible to reach any region of the globe, no matter how remote, without resorting to strategic aviation.

The greatest achievement of Soviet science and technology in the field of rocketry was the launching, on 4 October 1957, of the world's first artificial satellite.

In its launching, this satellite, which had the shape of a sphere 58 cm in diameter and weighing 83.6 kg, received its 8000-m/sec orbital velocity and its 900-km maximum distance from the earth's surface from a rocket carrier; it completed one revolution about the earth in 1 hour 35 minutes.

This satellite completed 1400 revolutions about the earth and traveled about 60 million kilometers.

This successful launching of a man-made Earth satellite was the greatest contribution ever made to the treasury of the world's science and culture.

The second artificial Earth satellite was launched in the Soviet

Union on 3 November 1957 (as part of the program of the International Geophysical Year).

This Earth satellite was the last stage of a rocket vehicle and carried a container for scientific apparatus and an experimental animal (a dog) and weighed a total of 508.3 kg.

The initial orbital characteristics of the satellite were as follows: shape, elliptical; altitude at apogee, 1671 km; inclination of orbit to equator, $65^{\circ}17'$; period of revolution, 103.74 min. The orbital velocity of the satellite at launch was 8000 m/sec.

This satellite completed about 2370 revolutions around the earth, and traveled a distance of 100 million kilometers.

With the successful launching of an artificial Earth satellite carrying various scientific apparatus and an experimental animal, Soviet scientists expanded their investigation of cosmic space and the upper layers of the atmosphere.

On 15 May 1958, a third Earth satellite — the largest in the world — was launched in the Soviet Union. It weighed 1327 kg. It was 3.57 m long and had a largest diameter of 1.73 m.

The initial orbital characteristics of this satellite were as follows: shape, elliptical; altitude at apogee, 1880 km; inclination of orbit to equator, 65°; period of revolution, 105.95 min.

This satellite continues in its motion about the earth. On 15 February 1960, it had completed 9221 revolutions.

The next magnificent attainment of Soviet rocket science and engineering was the creation of a multi-stage rocket the last stage of which was capable of reaching the second cosmic [parabolic] velocity (11.2 km/sec), thus making interplanetary flight possible. Such a cosmic rocket was successfully launched toward the moon in the USSR on 2 January 1959.

The last stage of the rocket, which weighed 1472 kg without fuel, was equipped with a special container for measuring apparatus and radio transmitters to carry out scientific investigations. The total weight of the scientific and measurement apparatus with their power sources and container came to 361.3 kg.

Thirty-four hours after blastoff, this cosmic rocket flew past the moon (at a distance of about 6-7 thousand kilometers) and was 597,000 kilometers from the earth in the 62nd hour of its flight, at 10 hours on 5 January. Having overcome the attraction of the earth and the moon, it emerged into its orbit about the sun.

Dependable radio contact between the rocket and the earth was maintained for 62 hours, in accordance with the program.

Evaluation of the measurement data obtained made it possible to establish that the cosmic rocket finally entered the periodic orbit of an artificial planet of the solar system on 7-8 January 1959. The largest diameter of the rocket's orbit, which lay between those of the earth and Mars, was found to be 343.6 million kilometers. The rocket's period of revolution about the sun is 15 months (450 earth days).

The launching of the cosmic rocket, which had become for all eternity the first artificial planet of our solar system, opened the era of interplanetary flight and demonstrated to the entire world the outstanding achievements of Soviet science and engineering.

Another historical first was scored in the period from 12 through 14 September 1959 with the successful launching of a second multiple-stage Soviet cosmic rocket to the moon; this device delivered to its surface a pennant with the coat of arms of the Soviet Union.

Valuable scientific data which opened up a new stage in the exploration and conquest of the cosmos by man were transmitted from the rocket and received on the earth during this flight.

The landing of a Soviet rocket on the moon represents an out-standing success for Soviet science and engineering and marks the beginning of flight from the earth to other planets.

On 4 October 1959, a third multiple-stage cosmic rocket was launched in the USSR to continue the investigation of cosmic space and photograph that side of the moon which is invisible from the earth.

For this purpose, an interplanetary space station that could be controlled automatically from the earth was designed, built, and placed by a rocket in a previously calculated orbit around the moon. As planned, this station passed at a distance of several thousand kilometers from the moon, photographed its other side on 18 October, and transmitted these images to the earth on command by means of a special radio communications system.

This rocket launching to the moon, which resulted in the acquisition of extremely valuable and hitherto unavailable scientific data concerning cosmic space, including the photographs of that side of the moon which is invisible from the earth, required solution of a number of highly diverse and complex problems by Soviet experts. These attainments of Soviet science and engineering have shaken the entire world and have been an occasion for enormous delight for all progressive humanity.

On the basis of the progress made in the Soviet Union in investigating cosmic space by means of ballistic rockets, Soviet scientists and designers are working in accordance with a scientific-research program on the creation of a more powerful rocket to be used in launching heavy Earth satellites and making space flights to other planets of the solar system.

On 8 January 1960, TASS reported that this rocket would be launched into the central part of the Pacific Ocean without its last stage during the first few months of 1960, as part of the development of a high-accuracy rocket of this type. The area in which the stage [sic] of the rocket was expected to fall was indicated in this report.

The first launching of this rocket took place in the evening of 20 January 1960.

At 20 hours 15 minutes Moscow time on the same day, the next-to-last stage of this rocket, having moved exactly along the specified trajectory with a dummy of the last stage and having developed a speed better than 26 thousand kilometers/hour, reached the specified area in the equatorial Pacific Ocean, which was 12.5 thousand kilometers across the earth's surface from the point of launching.

The last-stage dummy, which was adapted to pass through the dense layers of the atmosphere, reached the surface of the sea near the calculated impact point.

Special ships of the Soviet fleet waiting in the rocket's expected area of impact made valuable telemetric measurements on the descending branch of the flight trajectory.

The dummy of the rocket's last stage was observed during its flight through the atmosphere and was pinpointed where it dropped into the water by radar, optical, and sonar stations on the ships.

It was established by the measurements carried out that the rocket's impact point was less than 2 km from the predicted position, so that the high accuracy of the rocket's control system was confirmed.

The rocket was launched exactly at the specified time. The rocket's flight and the operation of all of its stages took place in conformity with the predetermined program. The measurement systems and facilities on board the rocket transmitted the necessary data to ground and ship-borne stations over the entire flight path.

A second such rocket was launched for the same purpose in the evening of 31 January 1960.

The last stage of the rocket with its dummy reached the specified area in the equatorial Pacific Ocean at 19 hours 58 minutes Moscow time.

The last-stage dummy was again observed during flight through the atmosphere and was pinpointed on impact with the water by ship-borne radar, optical, and sonar stations.

The measurement data again confirmed the high precision with which the rocket's flight was controlled.

The successful launchings of this powerful Soviet multiple-stage ballistic rocket have further advanced Soviet science along the path to the conquest of cosmic space and study of the solar system.

SECTION 2. DIRECTIONS IN THE DEVELOPMENT OF ZhRD

Analysis of the designs of existing ZhRD enable us to establish the basic directions being taken in the development of this branch of engineering.

Oxygen-, nitric-acid-, and hydrogen-peroxide engines were developed during the Second World War and used in missiles, aerial torpedos, and aircraft.

The oxygen-carrying rocket engines used at that time for long-range missiles developed absolute thrusts no greater than 26 tons, worked on fuels with relatively low calorific values (75% ethyl alcohol and liquid oxygen) with low combustion-chamber gas pressures (16 atmospheres absolute) and low thrusts per liter (about 60-80 kg per liter of volume) and, as a result, developed low specific thrusts [of the order of 200 kg of thrust/(kg of fuel/sec)].

The ZhRD of the A-4 long-range missile (Figs. 1.1 and 1.2) can be cited as an example of an oxygen engine of this type.

A tank for 75%-by-weight ethyl alcohol (the fuel) and a tank for liquid oxygen (the oxidizer) are located in the middle of the rocket. The 4600-liter fuel tank weighs 76 kg; it holds 3900 kg of fuel. The oxidizer tank has a capacity of 4470 liters and weighs 120 kg; 4900 kg of oxidizer are pumped into it.

The engine, which weighs 930 kg, is mounted inside the tail section of the missile.

The engine develops a surface thrust of 25-26 tons for 60-70 sec and consumes about 125 kg of fuel per second during this process. The basic elements of the engine are the welded pear-shaped combustion chamber (see Fig. 1.3), and the turbopump set and gas generator of the fuel-supply system. The gas mixture developed in the gas generator from 80%-by-weight hydrogen peroxide with the aid of a liquid catalyst (sodium permanganate) is used to drive the turbine of the pump set. The greatest outside diameter of the engine's chamber is 1190 mm, and it is 2020 mm long. The chamber weighs 422 kg.

Figure 1.4 shows an A-4 missile at takeoff.

During the Second World War, nitric-acid reaction engines were

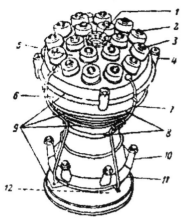

External appearance of chamber of A-4 engine. 1) Flange for attachment of main alcohol valve; 2) oxygen nozzle; 3) pre-combustion chamber; 4) mounts; 5) chamber head; 6) combustion chamber; 7) cooling-system pipeline; 8) corrugations to offset thermal expansion of chamber outer shell; 9) rings for alcohol supply to chamber; 10) connecting pieces for delivery of ethyl alcohol into space between tanks; 11) fuel-collector ring; 12) boundary of cooled section of nozzle.

built with thrusts ranging from 300 to 8000 kg and specific thrusts of about 190-200 kg of thrust/(kg of fuel/sec).

The S-2 liquid-fueled engine of the "Wasserfall" radio-controlled antiaircraft missile, which developed a surface thrust of 8 tons (Fig. 1.5) can serve as an example of a nitric-acid engine.

This missile was also developed in Germany and reached the flight-testing stage during the Second World War. In external appearance, it resembles the A-4 missile, but it is considerably simpler and has smaller over-all dimensions. Its most significant departure from the A-4 consists in the fact that it is equipped with four small, broad wings. It was the largest of all the guided antiaircraft missiles developed during this period of time for defense against enemy aircraft flying as high as 20 kilometers. Its body had a maximum diameter of 880 mm and a length of 7835 mm, its wing span was 1875 mm, and its tailfin span was 2510 mm.

The missile's warhead weighed 145 kg, of which 125 kg was explosive. The total blastoff weight of the rocket was 3245 kg.

This rocket's engine consists basically of an elliptical combustion chamber and the gas-bottle fuel-supply system. The combustion chamber (welded from sheet steel and integral with the nozzle) has a flat removable head with sprayers for the fuel components. The chamber is cooled by the oxidizer.

The engine is powered by components that ignite spontaneously on mixing: Tonka-841 compound (fuel) and M-10 compound (oxidizer). Compound 841 consists of 12% Optol, 20% benzene, 15% xylene, 30% of vinyl-ethyl ether, and 23% of aniline, while the M-10 compound consists of 90% of 98%-by-weight nitric acid and 10% of 96% sulfuric acid. One tank holds 335 kg of fuel and the other 1480 kg of oxidizer.

The components are fed into the combustion chamber of the engine under a pressure of 25 atmospheres, with the combustion chamber at 22.5 atmospheres. The compressed-nitrogen pressure in the tanks of the fuel-supply system is 300 atmospheres. The engine operates for 43 sec. The chamber weighs 43 kg, and the dry weight of the entire engine is 150 kg.

Work on the design of this engine was begun in the middle of 1941, and its first phase was completed in 1943. About 20 trial launchings of the missile had been carried out during this same span of time. The results of flight tests with the missile indicated the necessity of further work on it.

Series-production and experimental models of liquid-fueled rocket engines were designed and built during the Second World War for use in aerial torpedos and aircraft; these included "cold-type" hydrogen-peroxide engines.

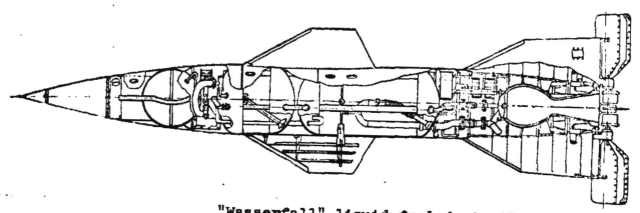

"Wasserfall" liquid-fueled missile.

The hydrogen-peroxide engines of this type which were constructed in this period of time and somewhat later developed small specific thrusts [about 100-110 kg of thrust/(kg of fuel/sec)], and were therefore not developed and used further.

The Type 109-507 aerial-torpedo ZhRD (Walther), which develops a thrust of 550-590 kg for 10-13 sec (Fig. 1.6) may serve as an example of the hydrogen-peroxide engine.

The 80% hydrogen peroxide used as a fuel in this engine is decomposed by a liquid catalyst (23%-by-weight sodium permanganate). The working components are fed into the reaction chamber by compressed air. The initial air pressure in the bottle is 150 atmospheres, and the working components are delivered at a pressure of 30-35 atmospheres. The reaction-chamber pressure is 22 atmospheres, and the temperature of the vapor gas formed in it reaches 420-440°C.

The engine can work until the hydrogen-peroxide supply in its tank is fully exhausted.

The cowling length of the engine is 2210 mm, and its largest diameter is 330 mm. Its dry weight is 17.5 kg, and when supplied with the working substances (hydrogen peroxide, aqueous solution of sodium permanganate, and compressed air), the engine weighs 143 kg.

All of the hydrogen-peroxide engine types developed in Germany by the firm Walther were used by the Germans during the Second World War as main and auxiliary engines for aircraft, aerial torpedos, and submarine torpedos.

However, despite the fact that hydrogen peroxide has no advantages over nitric acid, hydrogen-peroxide engines were used on a broader scale during the Second World War than nitric-acid engines. This is accounted for by the fact that they were brought to the operational state earlier.

The mastery of hydrogen-peroxide engines developing large absolute thrusts was paralleled by research in a number of countries toward the creation of oxygen-, nitric-acid-, and other engine types for high-powered long-range rocket missiles and antiaircraft-defense missiles.

The postwar development of reaction-engine technique required the creation of ZhRD developing large thrusts with minimal specific weights and specific fuel consumptions.

Solution of this problem required taking the course of increasing combustion-chamber dimensions, forcing the fuel-combustion process in it to increased pressures and temperatures, increasing the engine's fuel-consumption rate and per-liter thrust, and using more efficient fuels.

At the present time, the development of ZhRD is moving toward perfection of the designs of individual modifications and development of new high-thrust engine designs to use high-efficiency fuels and develop high specific thrusts. The combustion temperatures of some of these fuels may reach $4000°C$, and the gas-discharge velocity through the nozzle may run as high as 3000-4000 m/sec.[*]

The fuels in use at the present time include Tonka-250,[**] kerosene, ethyl alcohol diluted with water, etc.; aniline, Diesel fuel, furfuryl alcohol, etc. are in use abroad.

The following fuels are receiving a great deal of attention at the present time: hydrazine and fuels based on it (methyl hydrazine, dimethyl hydrazine), diethylamine, fuel mixtures based on kerosene and pyrolysis oil, etc. Liquid fluorine, oxygen fluoride, ozone-enriched oxygen, etc. are of great interest as oxidizers.

At the present stage in their development, long-range missiles require:

1) a sharp increase in the absolute thrust of the engines up to several hundred tons for a single unit;

---

[*]Ekspress-informatsiya AN SSSR, No. 8, RT-22, 1958.
[**]G. Mebus [sic], Raschet raketnykh dvigateley (Design of Rocket Engines), IL (Foreign Literature Press), 1959.

2) an increase in the specific thrust of the engines up to many tens of tons per unit;

3) a further reduction of engine weight per unit of thrust developed;

4) the creation of multiple-chamber engines with automatic operation of all units.

As science develops the field of liquid-fueled rocket engines, increasing amounts of attention are being concentrated on such problems as the reliability of the designs, simplicity of control, etc.

Great difficulties are encountered in the development of high-power engines. The inordinate intensity of the combustion process in the engine chamber, the high gas velocities, pressures, and temperatures, which are considerably higher than any values hitherto attained in chemically fueled machines, the enormous per-second fuel flow rate, which requires servicing by feeder units developing several tens of thousands of horsepower, the power of the engine itself, which may run to tens of millions of horsepower — all this renders particularly acute the problems of chamber cooling, making the chambers explosion-proof, starting the engine and controlling its operation, perfection of the TNA* and methods of supplying its turbine, the development of new high-efficiency fuels, high-strength welding steels, and many other extremely complex problems.

The most complex problem encountered in designing high-thrust engines is that of ensuring stable operating modes. This is due to the fact that difficulties are encountered with large engines of the A-4 type in connection with the necessity of making the inner chamber shell quite thick (over 5 mm). During operation, large temperature gradients arise in this shell, with the resulting complex distribution of thermal stresses across its thickness, the latter reducing its strength margin considerably because of the difficulty of arranging suitable cooling for such a thick shell, which has a large coefficient of thermal conductivity (materials possessing high thermal-conductivity coefficients frequently have relatively low strengths).

In a number of cases, this effect is further complicated by elastic instability of the shell as a result of the high pressure exerted on it by the cooling fluid and the thrust from the nozzle end. Moreover, high-thrust engines have large nozzles, which make up the basic part of the engine chamber's weight. And the use of monobloc engine designs involves great difficulty in synchronizing and controlling the operation of the block.

The development of high-thrust oxygen engines for SDD* is being paralleled in many countries by work on the creation of nitric-acid engines for these purposes (ZhRD "Erlikon" [sic] 54).

The development of high-powered liquid-fueled rocket engines has been accompanied by problems of controlling them, foremost among which are:

1) maintaining thrust and chamber pressure constant or varying them in accordance with a specified program;

2) maintaining the specified proportions between the fuel components, i.e., the weight ratio of oxidizer to fuel;

3) maintenance of dynamic stability in the fuel-supply system and the fuel-combustion process in the engine chamber;

4) programming the pressure variation in the combustion chamber and the proportions between the fuel components during static testing

of the engine;

5) compensation of periodic or constant errors stemming from tolerances in production of the engine.

Regulation of the fuel-component proportions during the engine's operation is necessary for maintenance of optimum specific thrust, which prolongs the engine's operation time and reduces the final mass of the missile, thereby increasing its range.

For ballistic missiles, the thrust must either be varied in accordance with a specified program dictated by the acceleration permissible or held constant. The latter case occurs only in the absence of aerodynamic drag and may therefore be used in the last stages of multiple-stage missiles. In other cases, optimum programming of the thrust value is required for attainment of maximum terminal velocity. This also applies to winged missiles.

One important problem encountered even when automatic-control systems are used, is that of ensuring dynamic stability of the fuel-combustion process in the chamber — a process strongly influenced by the time delay and nonlinearity of the combustion process, the rigidity of the design, including the fuel tanks and lines, the compressibility of the fuel components, and aerodynamic forces in addition to the shape of the chamber and its head and the chemical and physical properties of the fuel components used. Despite the availability of a number of studies devoted to theoretical investigation of this problem, the struggle with high- and low-frequency gas-pressure fluctuations in the combustion chamber is still being carried on by empirical methods.

Liquid-fueled rocket engines developing thrusts measured in hundreds of tons* were required in connection with the development of cosmic and super-long-range missiles in recent years, as well as missiles used in launching large Earth satellites. Nuclear energy will obviously come into use in the future for high-thrust engines.

Chapter 2

The basic element of liquid-fueled long-range, antiaircraft, and other rocket missiles is their engine — the ZhRD. The range of the missiles, other things being equal, depends to a large measure on the perfection of the engine's design, the type of fuel components used in it, and also its design and operational characteristics.

To evaluate the operational properties of existing types of ZhRD and to choose an engine as required by the tactical and technical conditions, at first one must become acquainted with some concepts, definitions, basic data and characteristics of the ZhRD.

In this chapter, the classification of existing ZhRD is stated, and their general characteristics are shown, permitting the ascertainment of their virtues and shortcomings, comparison of their qualities, and establishment of expedient fields of application for this or some other type of engine. Tables of the numerical values of basic parameters and brief data concerning the design particulars of existing ZhRD are included. Analyses of the basic factors affecting range and of the requirements for weapon engines are also shown. At the same time, other problems concerning the theory and bases of designing ZhRD are also considered.

SECTION 1. BASIC ZhRD DESIGN ELEMENTS

A heat engine is referred to as a reaction engine if it develops thrust because of the resultant of gasdynamic forces on the engine combustion chamber (as fuel is spent) and because of the discharge of the products of combustion through the nozzle into the atmosphere.

A reaction engine using liquid fuel for its operation is called a liquid-fueled rocket engine.

The basic design elements of a ZhRD are generally as follows.

1. The engine chamber, in which fuel combustion is accomplished and the heat energy of the gases is changed to kinetic energy in the outflowing stream, as a result of which thrust is developed.

The engine chamber consists of a head, combustion chamber, and nozzle. The head of the engine chamber serves for atomization of the fuel components, which are fed into the combustion chamber in a fixed weight relationship.

Spontaneous blending, heating, evaporation and combustion of the fuel occurs in the combustion chamber of the engine.

In the engine chamber nozzle, as the products of fuel combustion flow out into the surrounding medium, their heat energy is changed into kinetic energy. These processes have an enormous influence on the economy of operation and thrust characteristics of the engine.

The processes in the combustion chamber and in the engine nozzle are closely related to each other. The amount of chemical energy which may be liberated in the nozzle as the result of fuel combustion and recombination of gas molecules depends on the stage of completeness of the process of combustion in the combustion chamber.

2. The engine propellant-feed system, ordinarily consisting of one, two, or several fuel tanks, a mechanism for forced delivery of the fuel components to the engine chamber, a source of energy for starting this mechanism, communication systems and fittings (pipelines, valves, vents, flow-meter washers, and the like), which, in totality, provide

--

for normal launching, operating regime, and stopping of the engine.

In some cases, the fuel tanks are not included among the engine elements, but are part of the vehicle.

The engine may have such fuel-feed systems and controls as to permit it to be started and stopped if necessary.

3. <u>The engine ignition system</u>, which is a device for igniting the fuel when the ZhRD is launched.

In some types of ZhRD, the ignition system is not structurally connected with the engine chamber, and sometimes is even entirely lacking, if self-igniting fuel components are used.

4. <u>The engine thrust frame</u>, or other means of fastening engine assemblies to one another, and transmitting thrust to the weapon.

There are autonomous or on-board methods of feeding the engine units with compressed gases. For autonomous feeding, sometimes an air flask with a self-contained unit of reducers, pipelines, fittings, and valves which are necessary to create a pressure head in the fuel tanks and for other purposes, is used.

The rigid requirements demanded of an engine because of the great concentration of energy in the fuels used, the complexity of the physicochemical processes proceeding in it, and safety requirements during operation have led to the fact that contemporary ZhRD, in many cases, are, in their design relationships, extremely complex thrust frames with a highly developed automatic control system.

The aspiration for maximum automation of the engine's operation can be explained by the fundamental features of the ZhRD. In these engines, within a short period of time, it is necessary to carry out all operations which are required for dependable fuel ignition, increasing its delivery to the combustion chamber up to nominal value, maintaining this rate of flow constant or changing it in accordance with the programmed operation of the engine, and, finally, stopping the engine at the proper instant.

One must not forget that the fuel which is fed into the combustion chamber is an explosive fuel mixture. Considering the great expenditure of fuel per second in the engine, it is clear that the smallest disruption of the regularity of operation of the feed system or delay at the instant of ignition of the fuel may lead to its concentration in the combustion chamber and, consequently, to a sudden ignition with a sharp rise of pressure in the combustion chamber to a great quantity, and even, as a consequence of this, to explosion of the engine. The same thing may occur when fuel combustion is accidentally interrupted or a fuel is again fed into the hot combustion chamber after the engine is stopped. In this case, the ignition of the fuel by the hot surface of the chamber also may lead to explosion of the engine.

From what has been said it is clear that the fuel-feed system has to function without fail. In practice, this is attained by automation and the blocking of the engine fuel-feed system. By an automatic blocking system one understands the creation of such conditions as to prevent the following operation to be performed in an engine feed system from being carried out until such time as the previous operation has been accomplished.

The automation of the fuel-feed system in contemporary ZhRD is being carried to such a degree of perfection that all operations in launching, placing the engine on the chosen operating mode, and stopping it are carried out by delivering only one command to the engine.

Such a high level of ZhRD automation is also necessary because of the fact that they are usually installed in pilotless aircraft.

The missile's purpose determines its type of engine, thrust, and operation time, which in turn influences chamber dimensions, fuel-tank volume, and choice of fuel components.

SECTION 2. CLASSIFICATIONS OF EXISTING ZhRD

Existing liquid-fueled rocket engines are extremely diversified in their designs, operating characteristics, and other features. This is explained:

1) by the great diversity of fuels used in them;

2) by the purpose of this or that type of engine, which determines the magnitude of its thrust, program, and operation time;

3) by the features of the process involved in converting the chemical energy of the fuel, in the engine, into the kinetic energy of the gas stream, from the nozzle outlet to the surrounding medium;

4) by economic, production, and similar considerations, as well as by the features encountered in engine development trends at various design bureaus.

The field of application and the type of fuel used have the greatest influence on the design of a ZhRD.

To elucidate the virtues and shortcomings of these or those types and designs of engines, to establish expedient fields of application for them, and to study their design and operational details, it is expedient to divide existing ZhRD according to the following most characteristic features.

1. <u>According to engine designation:</u>

a) <u>sustainer or basic,</u> when each engine is basic in the given vehicle and operates during the entire flight or its greater part;

b) <u>booster,</u> used to facilitate launching of a weapon which has a sustaining engine;

c) _verniers_, used in a weapon during flight in addition to the basic engine for the purpose of providing a short increase in thrust and speed to the weapon.

Liquid-fueled rocket verniers are often used in aviation; they often have pump-fed fuel systems with mechanical drive from the main engine of the aircraft. The verniers may be started many times during the aircraft's flight.

Besides this, ZhRD may be intended for one-time (one-shot) operation, that is, for use during only one flight after installation in the vehicle, or multi-time (multi-shot) operation, that is, for use in several flights.

2. _According to the type of fuel used_, ZhRD are divided into engines operating with hypergolic and nonhypergolic fuel components. The various properties of fuels impose specific features on the design of the engine.

The selection of liquid-fuel components for a given engine is generally based on the methods of application, availability of components, characteristics, properties, and similar factors.

ZhRD fuels may be monopropellant, when only one liquid fuel component (isopropyl nitrate, nitromethane, hydrazine, and others) is used, and bipropellant, when two liquid fuel components — the fuel and oxidizer — are used. Tripropellant fuels also exist.

At the present time, bipropellant engines are most widely used.

3. _According to the type of oxidizer used for the fuel_, engines are divided into:

a) _oxygen_, using liquid oxygen or its allotropic modifications and compounds with the fuel elements as an oxidizer;

b) _nitric acid_, using, as the oxidizer, nitric acid and oxidizers which are derivatives of the nitric acid or which contain nitric acid;

c) <u>hydrogen peroxide</u>, using hydrogen peroxide with a liquid or solid catalyst;

d) <u>fluoride</u>, using fluorine, fluorides of oxygen, and other fluorine-containing compounds as oxidizers;

e) <u>chlorine</u>, using chlorine, oxides of chlorine, or other oxidizers containing oxides of chlorine or their derivatives.

Complex oxidizers, containing, in various combinations, molecules of oxygen, nitrogen, chlorine, and fluorine, as well as solutions of several oxides, acids, and other components in each other are also known. *

Engines which operate on suspensions of metals and metalloids, with liquid fuels, are also possible. **

The classification of engines according to the type of oxidizer used is most essential, since differences in oxidizer properties determine the design shape of the engines. There is no engine which can operate with several different oxidizers.

Every engine is developed for a previously determined oxidizer, and, as a rule, the design of one engine differs from the design of another because of the differences in the properties of the oxidizers used. The development of a ZhRD always begins with the choice of oxidizer and fuel for the engine.

4. <u>According to the method of propellant feed to the combustion chamber</u>, ZhRD are divided into the following classes:

a) with a pressure-fed system of fuel feed by means of:

— <u>gas pressure generators</u> (GAD),*** that is, the pressure of a cold gas, ordinarily air (VAD),* fed to the fuel tanks from a special flask;

— solid propellant hot gas generator (PAD),** that is, the pressure of the hot powder gases, which form during the operation of the engine in a special chamber by the ignition of a powder charge;

— liquid gas pressure generator (ZhAD),*** that is, the pressure of the hot combustion products of self-igniting fuel components, formed during the operation of the engine in one general or in two special separate chambers (gas generators), installed on the upper end plates of the fuel tanks;

b) with a pump-fed system of fuel feed by means of:

— turbopump unit, that is, with delivery of fuel components from the tanks to the engine's combustion chamber by centrifugal pumps, which are started by a gas-vapor turbine, fed with gas vapor produced in a special gas generator from hydrogen peroxide, isopropyl nitrate, or hydrazine, from ignition of the basic fuel components, or from gas obtained from the engine combustion chamber;

— injectors, whose operation is based on the principle of using the kinetic energy developed by a gas when it expands in a special nozzle (the gas required for injector operation is obtained from the combustion chamber or produced in a special gas-vapor generator).

A pump-fed propellant-feed system with a gas-pressure generator (GAD) is often called a pressurized fuel-feed system.

5. According to the heat load, ZhRD are:

a) **of the "hot" type**, in which the fuel burns at a high temperature (about 2700-3600°C), and

b) **of the "cold" type**, in which decomposition of hydrogen peroxide occurs at a comparatively low temperature (about 320-480°C).

6. **According to the method of cooling the chamber**, ZhRD are divided into engines having:

a) **regenerative cooling**, which involves the fact, that one of the fuel components (or sometimes both components), before arriving in the combustion chamber, pass through the interliner space of the chamber, and thus cools the inner liner of the nozzle and the combustion chamber;

b) **sweat cooling**, in which the coolant flows from the interliner space into the chamber through small pores in the inner liner which is made of a special porous material; thus the chamber wall is cooled and, simultaneously, a vaporized gas film is formed on the inner surface, protecting the liner from excessive heating by hot gases;
gases;

c) **flowing water cooling**, ordinarily used in test stands.

It is also possible to cool engine chambers by circulating water, which is at the same time the working fluid for the turbine of the pump unit of the engine propellant feed system (closed-cycle regenerative engine cooling).

The heat release rate of the combustion chamber and nozzle have a basic influence on the choice of the appropriate method of engine cooling.

At the present time, the regenerative method of cooling is the most widespread, since it is most dependable and economical. In this case, the heat which is transferred from the inner liner to the coolant is returned to the combustion chamber.

For regenerative engine cooling, the fuel component which has the least corrosive properties, the highest values of heat capacity, thermal conductivity, and other characteristics useful for this purpose is ordinarily used. The oxidizer is ordinarily used for cooling the chambers of small-thrust engines, since the fuel in these is insufficient for dependable cooling.

7. According to the method of protecting the inner liner of the chamber from overheating (with regenerative cooling), ZhRD may be divided into engines:

a) with a gas fuel curtain, created on the head side of the chamber through peripheral low-capacity spray nozzles;

b) with film fuel curtains, created in the most intensely heated parts of the engine chamber; the fuel is fed to the inner liner surface through special apertures or slits in it, by which means the liquid flows along this surface in the direction of the gas stream; it is gradually heated and vaporized, and thus protects the liner against excessive heating;

c) with insulation of the liner gas surface against heat of the gas flow (ceramics, graphite, metallic oxides, and other substances may serve as insulators).

Protection of the chamber liner against overheating by a gas or film fuel curtain, like the sweating method of cooling, is ordinarily applied in those cases when, as a consequence of the high heat release rate in the engine chamber, it cannot be cooled by the most simple and economical manner, or when the use of the latter is accompanied with great difficulties under the particular conditions.

The coolant duct of a ZhRD chamber may be slotted, spiral, spiral slotted, and of other shapes.

The ring-shaped coolant duct is the simplest and cheapest from a design viewpoint.

There are single-liner and double-liner chambers. Engines of the "cold" type and uncooled engines of the "hot" type which are intended for operation for not longer than 5-15 seconds have single-liner chambers. Cooled engines of the "hot" type with a relatively long operating period have double-liner chambers.

The engine chamber cooling system must permit the coolant to remove local heat flows, which have maximum value near the critical nozzle section, from the inner chamber liner with permissible heating of this fluid in the duct.

8. <u>According to the number of combustion chambers, engines are divided into</u>:

a) <u>monochambered</u>, that is, having only one combustion chamber in their design, and

b) <u>multichambered</u>, that is, having in their design several combustion chambers, capable of operating, as desired, either simultaneously or separately for the purpose of changing the magnitude of the engine's thrust.

ZhRD chambers are fabricated of steel, copper and steel, aluminum, ceramics and steel, and from other materials.

There are cylindrical, conically tapering, elliptical, pear-shaped, spherical, and other shapes of ZhRD combustion chambers.

The type of fuel used, method of atomization, pressure in the combustion chamber, magnitude of thrust and operation time of the engine, technology of its manufacture, cost, and other factors influence the selection of an expedient shape for an engine combustion chamber.

The outlet portion of a ZhRD chamber nozzle is constructed either:

a) conical (the flare angle of the outlet section of

the nozzle usually varies from 25 to 35°) or

b) shaped (to obtain a flow of gases which is axial, or close to it, in the outlet section of the nozzle).

Nozzle-section height may be controlled or noncontrolled.

9. According to the method of atomizing fuel components, there are ZhRD chambers:

a) with jet atomization (resembling the "Wasserfall" guided antiaircraft missile's engine);

b) with flat-spray atomization (resembling the "Rheintochter" antiaircraft missile's engine);

c) with centrifugal atomization;

d) with prechamber atomization (resembling the A-4 engine).

Both monocomponent and bicomponent centrifugal nozzles are made.

In construction, welded and detachable chamber heads have flat, tent-shaped, spherical, and other shapes. In a combustion chamber of spherical shape, that part in which the fuel-atomizing devices are located is called the head.

10. According to the method of igniting the basic fuel components in launching, ZhRD are divided into engines:

a) with chemical ignition, that is, by means of fuel components, basic or launching, which are self-igniting upon contact;

b) with electrical ignition, that is, by means of electrical devices (electric spark plugs or arcs);

c) with pyrotechnical ignition, that is, by means of a pyroelectric cartridge (a flame formed by the burning of a propellant grain).

11. According to the magnitude of nominal thrust, one may conventionally divide ZhRD into engines:

a) of small thrust (of the order of 0.5-5 tons), intended for aerial torpedos of various purposes, small antiaircraft missiles, launching engines, and for main engines for aircraft;

b) _of medium thrust_ (of the order of 5-25 tons), intended for large antiaircraft missiles and aircraft, small long-range missiles, high-performance and super-hypersonic fighter, interceptor, and reconnaissance aircraft, and

c) _of high thrust_ (over 25 tons), intended for large and extra long-range missiles.

ZhRD are produced to provide both for controlled and noncontrolled thrust during engine operation.

Control of the magnitude of thrust of a _ZhRD_ is accomplished:

a) by changing the fuel rate of flow per second in the combustion chamber by means of changing feed pressure and

b) switching on or switching off part of the fuel nozzles, or separate chambers (if the engine is multichambered).

If the fuel tanks are among the engine elements, _ZhRD_ differ as follows:

a) _with sequential tank arrangement_ (one tank arranged behind the other, along one axis);

b) _with concentric tank arrangement_ (one tank placed inside the other).

Fuel tanks may also be subdivided as follows:

a) _airframe_, which are both missile body and load carrier, and _nonairframe_, which are placed in the missile body and carry only the static load of the working components which are located in them;

b) _relieved_ of working gas pressure (with turbopump fuel delivery) and _loaded_ with working gas pressure (with a pressure-fed fuel feed system with a _VAD_, _PAD_, or _ZhAD_);

c) _"cold,"_ intended for liquid oxygen; _hot_, from which a fuel component is injected by a hot gas, and _normal_, from which a fuel component is injected into the combustion chamber by a cold gas.

12. <u>According to the engine's connection with the weapon</u>, ZhRD may be distinguished as follows:

a) <u>constructions which are independent</u> of the weapon (the engine is suspended from the weapon or installed in it);

b) <u>constructions, which are part</u> of the weapon (the engines of antiaircraft, short-range, and long-range missiles).

SECTION 3. BASIC ZhRD PARAMETERS

The basic parameters characterizing the flight and operational properties of a ZhRD of any type and design are as follows.

1. <u>Absolute or total thrust</u>, developed by the engine in operation at a rated operating regime, P_Σ (kg, ton).

The absolute thrust of the engine chamber, relative to the volume of the combustion chamber V_k,* is called volumetric thrust:

$$P_1 = P/V_k \text{ kg thrust/liter.**} \tag{2.1}$$

2. <u>Specific engine thrust</u>, that is, the thrust relative to the sum of the fuel rate of flow per second in the engine:

$$P_{ud\Sigma} = P_\Sigma/G_\Sigma \frac{\text{kg thrust}}{\text{kg fuel/sec}} \tag{2.2}$$

or

$$P'_{ud} = \gamma_t P_{ud} \frac{\text{kg thrust}}{\text{liter fuel/sec}},$$

where P_Σ is the sum of the engine thrust in kg; G_Σ is the total rate of fuel flow in the engine, consisting, in the aggregate, of rate of fuel flow G_s in the combustion chamber, and G'_s in the gas generator of the fuel supply system, and the rate of fuel flow G_{zav}*** in forming the protective curtain around the surface of the burner liner,

---

*[$V_к = V_k = V_{kamera} = V_{chamber}$.]
**[$P_л = P_1 = P_{litrovaya} = P_{volumetric}$.]
***[$G_{заs} = G_{zav} = G_{zaves} = G_{curtain}$.]

in kg/sec; γ_t* is specific gravity of the fuel in kg/liter; P_{ud}** is specific chamber thrust.

In existing ZhRD, it is possible to determine specific thrust from the thrust value and the rates of fuel flow into the combustion chamber, obtained by test-stand measurements.

Absolute thrust is not characteristic of the degree of perfection in ZhRD operation. An operational characteristic of an engine's operation is its specific thrust, sometimes called the specific impulse.

The greater the engine's specific thrust, the less the consumption of fuel per second will be at a given quantity of absolute thrust, and, consequently, the smaller the required supply of fuel in the tanks will be for any given period of operation, dimensions, and weight of the engine.

The greater the specific thrust of a ZhRD in operation on a given fuel, other things being equal, the better established the operating cycle in the engine and the more perfect the engine from a design viewpoint.

Absolute and specific engine-chamber thrust are constant quantities, with a given regime, for operations in space.

The required thrust of a ZhRD is determined in the design of the given weapon (depending on range, quantity of payload, and other factors).

3. The specific fuel consumption in the engine, that is, fuel consumption referred to a unit of engine thrust per second or per hour:

$$C_{ud} = \frac{1}{P_{ud}} \frac{kg\ fuel/sec}{kg\ thrust} = \frac{1}{P_{ud}} 3600 \frac{kg\ fuel/hour}{kg\ thrust}.*** \qquad (2.3)$$

*[$\gamma_T = \gamma_t = \gamma_{toplivo} = \gamma_{fuel}$.]
**[$P_{yd} = P_{ud} = P_{udel'nyy} = P_{specific}$.]
***[$C_{yd} = C_{ud} = C_{udel'nyy} = C_{specific}$.]

The specific fuel consumption, equally with the specific thrust, is an important operating characteristic of the engine, since the greater its value, other conditions being equal, the greater the range and operating period of a weapon will be with the given engine.

The quantity C_{ud} depends on the pressure in the combustion chamber, the type of fuel used, and the establishment of the operating cycle in the engine chamber.

4. <u>The specific weight of the engine</u>, that is, the weight, referred to unit of engine thrust:

$$\gamma_{dv} = G_{dv}/P_\Sigma \text{ kg/kg thrust,}* \qquad (2.4)$$

where G_{dv} is the dry weight of the engine in kg, and γ_{dv} is the specific weight of the engine.

By the dry weight of the engine is meant the weight of its entire structure when the tanks are not filled with operating components (fuel, gas, and the like).

At a given magnitude of absolute thrust, specific weight determines the total dry weight of the engine, which affects the weapon's characteristics to a considerable measure.

5. <u>The engine's frontal thrust</u>, that is, the thrust relative to 1 cm^2 of the greatest cross-section of the engine:

$$P_{lob} = P_\Sigma/F_{dv} \text{ kg thrust/cm}^2,** \qquad (2.5)$$

where F_{dv} is the greatest area of cross-section of the engine in cm^2.

The specific frontal thrust is very important when evaluating the aerodynamic qualities of the engine, since the greater its value at a given thrust, the less the greatest cross-section of the weapon may be

---

*[$\gamma_{дв} = \gamma_{dv} = \gamma_{dvigatel'} = \gamma_{engine};$
 $G_{дв} = G_{dv} = G_{dvigatel'} = G_{engine}.$]
**[$P_{лоб} = P_{lob} = P_{lobovaya} = P_{frontal};$
 $F_{дв} = F_{dv} = F_{dvigatel'} = F_{engine}.$]

with this engine. Existing engines have a P_{lob} = 1-4 kg thrust/cm^2.

To parameters characterizing the engine we should add the effective efficiency, i.e., a relationship between fuel components χ, combustion-chamber volume V_k, and chamber pressure p_k.*

Ordinarily, tne initial parameters of a ZhRD are given as its relative parameters in operation in a rated operating regime, reduced to standard atmospheric conditions.

Under the conditions mentioned, existing missile ZhRD (the "Vanguard" missile and others) have $P_{ud} \approx 200\text{-}240 \frac{\text{kg thrust}}{\text{kg fuel/sec}}$; $C_{ud} \approx 15\text{-}18 \frac{\text{kg fuel/hour}}{\text{kg thrust}}$ = 4.0-5.0 $\frac{\text{kg fuel/sec}}{\text{tons, thrust}}$; γ_{dv} = 10-40 kg/ton thrust.

The operation qualities of an engine are its adaptability and dependability of operation, lifetime, simplicity of maintenance, and others.

The engine parameters enumerated above facilitate evaluation of a given type of engine with respect to thrust force, economy of operation, external dimensions, and operational qualities, and determine, in the first approximation, the basic virtues and shortcomings of this or that engine; and establish rational fields for its application.

The basic parameters of an engine, which determine its qualities, depend on design particulars, operation modes of the engine itself, external conditions, and a number of other factors. The effective caloric value of the fuel and the pressure in the combustion chamber and in the nozzle outlet section have a substantial influence on specific thrust, specific fuel consumption, and efficiency of the engine. Therefore, to determine the basic qualities of a projected engine and properly choose its parameters, it is necessary to know the dependence of absolute and specific thrust, and of specific and per-second fuel con-

---

*[$p_к$ = p_k = p_{kamera} = $p_{chamber}$.]

sumption on flight altitude, pressure in the combustion chamber, and other factors.

SECTION 4. ADVANTAGES AND DISADVANTAGES OF ZhRD

At the present stage of development of reaction engineering, only reaction engines which operate by the chemical energy of decomposition and oxidation of fuel have been realized and are in practical use.

Such engines include reaction engines operating on solid fuel (solid-propellant or PRD*), on liquid fuel in an air atmosphere (VRD**), and the ZhRD, operating on liquid fuel; these engines are, at the present time, experiencing their greatest development and application.

Liquid-fueled rocket engines have the following advantages in comparison to air-breathing reaction motors:

1) they can develop thrust even in an airless space, since their operation does not depend on the surrounding medium (flying altitude is limited only by the supply of fuel in the tanks and the engine's specific thrust);

2) absolute thrust, in contrast to the air-breathing reaction motors, increases with ascent in altitude, and is little dependent on flying speed in the dense layers of the atmosphere;

3) the possibility of concentrating extremely high thrust (up to 1000 tons) in one engine at comparatively small dimensions and specific engine weight (from 0.01 to 0.040 kg/kg of thrust)***;

4) the total weight of the engine is less than the weight of any air-breathing engine of the same power;

5) the convenience of using the engine with a weapon as a result

---

*[ПРД = PRD = Porokhovoy Raketnyy Dvigatel' = Solid-Propellant Rocket Motor.]
**[ВРД= VRD = Vozdushno-Reaktivnyy Dvigatel' = Ram-Jet Engine.]
***Ekspress-informatsiya, AN SSSR, No. 20, ADS-78, 1958.

of the relatively small external dimensions;

6) the possibility of obtaining relatively great speeds and flying altitudes, which are not attainable for a VRD (the A-4 long-range missile has a flying speed of about 1520 m/sec for 5500 km/hr at an altitude of 35-37 km);

7) a special engine is not required for launching.

Like other types of engines, the ZhRD permits the accomplishment of multiple start-ups and regulation of thrust force by changing per-second fuel rate of flow in the combustion chamber.

The basic disadvantages of the ZhRD are:

1) the low economy of operation at low flying speeds (a great specific fuel consumption);

2) the short operating time of the engine, which is governed by the extremely high specific fuel consumption [up to 15-20 (kg of fuel per hour/kg of thrust)];

3) the insignificant operational line of the engine (from 2.5 sec up to 2 hours instead of 50-300 hours for a VRD and 100-350 hours for piston engines);

4) the payload of the weapon is decreased with increase of range by the increase of the supply of fuel in the tanks;

5) the necessity of having a liquid fuel oxidizer in the rocket aircraft, the handling of which is often attended with great difficulties.

The advantages and disadvantages of the ZhRD mentioned above have determined their fields of application.

However, in comparison with engines of other types, these virtues and shortcomings of a given ZhRD may be more fully elucidated only by inspection of the concrete conditions of its operation, such as thrust program, altitude and flying speed, and so forth.

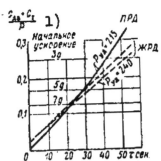

Fig. 2.1. Approximate comparison of weight characteristics (including fuel) of a PRD and ZhRD at various operation times. 1) Initial acceleration.

In some cases (with short operation times, small amount of thrust, or in special fields of application) the solid-propellant rocket engine (PRD) can successfully rival the ZhRD.

Among the virtues of the PRD there are, first of all, simplicity of its construction and method of loading the fuel, constant readiness for operation, reliability in operation, cheapness, less specific weight, less

TABLE 2.1

1) Тип двигателя	2) Максимально достижимая скорость полета км/час
3) Поршневые	600
4) Турбовинтовые	900
5) Турбореактивные	2 000
6) Прямоточные ВРД	6 000
7) Пороховые РД	28 000
8) Жидкостные РД	50 000
9) Термоатомные РД	100 000
10) Электрононные РД	500 000
11) Фотонные РД	$300 \cdot 10^6$

1) Type of engine; 2) maximum flying speed attained, km/hr; 3) piston; 4) turboprop; 5) turbojet; 6) ram-jet; 7) solid-propellant rocket engine; 8) liquid-propellant rocket engine; 9) nuclear-heated rocket engine; 10) electric-(system) rocket engines; 11) photon rocket engine.

dependence of the construction on the type of weapon, and widespread application in most diverse fields of contemporary engineering. Moreover, the operation of a solid-propellant engine does not depend on the maneuvering of the missile and its acceleration in flight.

Solid fuels (powders) have a somewhat smaller specific impulse than liquid fuels and are more expensive, but their comparative advantage lies in their greater specific weight. Powders, in comparison with liquid fuels, are more reliable and safe in operation. They can be better stored over extensive periods at any atmospheric temperature, they are not toxic, not aggressive in relation to metals, are ready for use at any moment, do not require any sort of prolonged and complex preparations before use, and so forth.

However, the weight characteristics of solid-propellant engines can be better, in comparison with the ZhRD, only when they are operated for short periods. When the engine operates for more than 20 seconds, the ZhRD has a smaller relationship of engine weight (including fuel as well) to the developed thrust than a PRD. Even with equal specific thrust [P_{ud} = 215 (kg of thrust/kg of fuel per second)], the weight characteristics of a PRD may be compared to those of a ZhRD only when time of operation is of the order of 26 sec (Fig. 2.1). When the time of operation is less, the PRD has better weight and thrust characteristics than the ZhRD, since they provide an increased thrust without proportional increase of the weight of the dry construction for a given weight of charge. Moreover, as a consequence of the lower relationship of weight to thrust, the PRD permits the attainment of greater missile accelerations.

Another basic reason for the limited application of solid-propellant rocket missiles in a number of cases is the high cost of powder, the large dimensions of the powder chamber for a considerable time of operation, and the appearance of an unstable, resonant combustion of powder in long chambers of small diameter, when the hot powder gases flow in the chamber with great speeds.

The tentative values of flying speeds attained in practice for aircraft with different types of engines are shown in Table 2.1.*

SECTION 5. FIELDS OF APPLICATION OF ZhRD

Liquid-fueled rocket engines have extremely extensive fields of application.

At the present time ZhRD are basically used for:

1) guided and nonguided missiles (antiaircraft, short and long range, cosmic, and others);

2) fighter aircraft, as the basic power plant, and bombing aircraft, as a booster engine, intended for short aircraft maneuvers, increasing horizontal and vertical flying speeds, and altitudes.

The ZhRD has a comparatively limited application in meteorological missiles, racing automobiles and motorcycles, and cargo trucks intended for operation on bad country and mountain roads (to facilitate motor-vehicle movement).

Unguided liquid-fueled rocket missiles (NRS**) are chiefly used as a tactical weapon for bombardment at short distances.

Guided liquid-fueled rocket missiles (UZhRS***) basically resemble unguided missiles in their external appearance and are guided during flight by means of special devices. These missiles may be used for strategic targets, as a means of PVO,**** and for carrying out scientific studies of outer space. Such missiles have extremely diverse purposes ("ground-ground," "ground-air," "ground-water," "air-air," and others).

Since the direction of flight is corrected by devices in these missiles, the accuracy of fire in this case considerably exceeds the accuracy of fire of unguided liquid-fueled and solid-propellant rocket missiles.

Flight range and altitude of single-stage UZhRS without booster motors is limited by the difficulties of constructing such missiles with great relationship of weight of dry construction and payload to gross take-off weight of the missile, and the comparatively low effectiveness of the fuels used.

Aircraft with a ZhRD as their main engines, in comparison with aircraft having engines of other types (VMG, TVRD, TRD,* and PRD), have advantages in regard to ceiling, flying speed, and rate of climb.

At the present time, an aircraft with this engine is the only vehicle which can accomplish a flight at exceptionally high altitudes and with very great speeds. However, the great rate of fuel flow per second in a ZhRD limits the range and duration of flight and other important characteristics of the aircraft. Moreover, methods of increasing the range of an aircraft with this engine are not promising. Therefore, it is expedient to use the ZhRD only for aircraft intended for short flights and high altitudes.

Rocket engines have created the possibility of conquering interplanetary space. However, the solution of this problem still requires enormous work along the lines of increasing the efficiency of the engine, using the more effective liquid fuels which are exploited with difficulty, constructing missiles with maximum relative fuel conservation, and with the use of multistage missile plans, etc.

Atomic reaction engines will be applied without limitations both for flights from the earth and beyond the earth, given the condition that the basic problems connected with their development are satisfactorily solved.

SECTION 6. REQUIREMENTS DEMANDED OF WEAPON ENGINES

Weapon engines demand the following basic requirements:

1) compactness and simplicity of construction, and cheapness of mass production and operation;

2) automation of operation and reliability of hermetic sealing of fuel-feed equipment;

3) small specific weight and dimensions for an engine of given thrust;

4) continuous combat readiness for launching within the temperature range of -40 to +50°C;

5) reliability of launching, operation, and stopping;

6) good response — stable thrust increase (for missile engines, the time required to reach an operating regime of 90% nominal thrust must not exceed 1-3 sec, counting from the instant at which the launching pulse is delivered);

7) high economy of operation (the greatest possible value of specific thrust);

8) the fuel combustion process in the engine chamber must be dynamically stable in a given thrust range, during the weapon's maneuvers and perturbations during flight;

9) constancy of developed thrust for the established operating regime;

10) possibility of stopping the engine before complete expenditure of the fuel supply in the tanks (if necessary);

11) possibility of prolonged storage of the engine, filled up with fuel, and transporting it by motor or railway transportation (this is especially important for the engines of antiaircraft missiles and aerial torpedos);

12) materials of the engine and its anticorrosive covering must permit its prolonged storage;

13) convenience of assembling the engine and the weapon, and others.

This list of requirements is equally applicable to expendable engines (antiaircraft and long-range guided missiles and others) and multi-shot engines (the aviation type).

Aircraft engines for multi-shot operation, boosters, and launching engines demand the following additional requirements.

a) For multi-shot engines:

1) possibility of starting it at any flight altitude in not more than 3 sec;

2) possibility of not less than 6-7 starts during flight;

3) regulation of thrust within the limits from maximum value to 0.1;

4) control of operation from one control unit;

5) operating life must be not less than 2 hours.

b) For boosters for main engines:

1) possibility of starting at any altitude in not more than 3 sec;

2) possibility of not less than 3-4 starts during flight;

3) regulation of thrust (if necessary) from maximum value to 0.3;

4) operating life of the combustion chamber must be not less than 1 hour.

c) For launching engines:

1) possibility of synchronized starting of several engines operating in parallel;

2) possibility of jettisoning to the earth by parachute after completion of operation.

Blastoff of unguided aircraft and ballistic missiles and transmitting the initial acceleration of flight to them are possible by means of the application of a powder charge in the combustion chamber of the sustaining ZhRD.

The above list of requirements, in general, is entirely complete, and, moreover, it characterizes the operational potentials of engines of various designations rather fully.

In each separate case, an engine may also demand other additional requirements.

In designing nitric-acid engines, it is necessary to consider additionally the extremely high toxicity and corrosive properties of nitric acid, and its continuous vaporization (fuming). The insulation of a nitric-acid engine and tanks in a weapon require a special method, which provides complete insulation of the engine from the other parts of the vehicle. Materials used for engines of this type must be corrosion-resistant to nitric acid.

SECTION 7. FACTORS AFFECTING VELOCITY AND RANGE OF MISSILES WITH ZhRD

Many various factors, of which the following are the most basic, affect a missile's velocity V_{kon}* at the end of the powered phase and range L:

1) the fuel-weight ratio in the missile:

$$a = G_t/G_0; ** \qquad (2.6)$$

---

*$[V_{кон} = V_{kon} = V_{konets} = V_{end}.]$
**$[G_т = G_t = G_{topliva} = G_{fuel}.]$

2) the missile's thrust-weight ratio:

$$b = P_0/G_0;$$
(2.7)

3) the engine's specific thrust:

$$P_{ud_0} = P_0/G_\Sigma,$$
(2.8)

where G_t is the supply of fuel in the missile's tanks before blastoff; G_0 is the total starting (initial) weight of the missile; P_0 is the engine's absolute thrust at missile blastoff; G_Σ is fuel rate of flow per second in the engine;

4) the angle of the missile's flight trajectory at the end of the powered phase relative to the vertical;

5) the aerodynamic shape of the missile, principle of its multiple stage design, and others.

The fuel-weight ratio a for the missile has the greatest effect on V_{kon} and L, but in practice it is limited by the structural features of the missile and the materials used.

In existing antiaircraft missiles a ~ 0.60-0.65, and in long-range missiles a ≈ 0.72-80.

The lower value of the coefficient a in antiaircraft missiles, compared with long-range ballistic missiles, can be explained by the fact that the antiaircraft missiles are more rugged and heavier since they are likely to encounter greater lateral acceleration during flight, and consequently greater loads.

Long-range missiles that are more highly perfected with respect to weight are possible, but at the present time it is possible to attain this only at great cost (by the application of multistage missiles, use of more efficient fuel, etc.).

At given values of P_{ud} and b, the quantities V_{kon} and L are directly proportional to a.

The missile's thrust-weight ratio b has the least influence on

V_{kon} and L relative to $\underline{a}$ and P_{ud}, and, moreover, has limited practical potential. When $\underline{b}$ is increased, flight range, at first, increases swiftly, but when $\underline{b}$ is greater than 10 it remains almost constant. With given values of $\underline{a}$ and P_{ud}, an increase of $\underline{b}$ increases the values of V_{kon} and L in direct proportion to $\underline{a}$. In practice, even when $\underline{a}$ is equal to 0.80 (a long-range missile), it is expedient to assume a value of $\underline{b}$ greater than 3, since to increase $\underline{b}$, it is necessary to increase the absolute engine thrust, and, consequently, the dimensions and weight of the engine and missile, which under certain conditions causes the gain due to increased $\underline{b}$ to vanish.

In existing antiaircraft missiles, $b \approx 2.5-6$ and higher, and in long-range missiles, $b \approx 1.8-2.5$.

At a lower thrust-weight ratio, the missile has bad blastoff properties, and at higher thrust-weight ratios, bad weight characteristics, which decreases the effective use of the missile.

The theoretical optimum of the missile's thrust-weight ratio depends basically upon its designation, type, degree of design perfection, and launching method used.

The greater the thrust-weight ratio of the missile, the greater its rate of climb and axial overload. One of the factors determining the upper limit of the missile's thrust-weight ratio is the axial overload permitted by operating conditions and durability of the missile's on-board guidance system devices.

The engine's specific thrust P_{ud} is the second most important factor after the coefficient $\underline{a}$, in increasing V_{kon} and L (the velocity V_{kon} is proportional in the first stage, and L in the second stage, to the quantity P_{ud}).

With a given value of $\underline{b}$, the influence of P_{ud} on V_{kon} and L is directly proportional to the factor $\underline{a}$. However, an increase of P_{ud} re-

TABLE 2.2

[1] Увеличение a при $P_{уд}=200$ ксек/кг, %	[2] Повышение L относительно дальности снаряда А-4 за счет увеличения a при $P_{уд}=200$ ксек/кг $L=350$ км принята за единицу	[3] Увеличение $P_{уд}$ для такого же повышения L, как и за счет роста a для $a=70\%$
		[4] $P_{уд_0}=200$ кгсек/кг
70	1	
85	2,3	350
90	4	350
95	6	400

1) Increase in a, at P_{ud_0} = 200 kg sec/kg, %;

2) increase in L, relative to range of A-4 missile, due to increase in a, at P_{ud_0} = 200 kg sec/kg, L = 350 km is the unit; 3) increase in P_{ud} for the same increase in L, as well as for the increase in a, at a = 70%; 4) P_{ud_0} = 200 kg sec/kg.

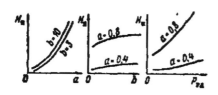

Fig. 2.2. Flight altitude H of missile, as function of values of a, b, and P_{ud}.

quires raising the pressure in the engine combustion chamber, the use of fuels with higher heating values at high combustion temperatures and, thus a structurally more complex cooling system for both combustion chamber and nozzle; also, other measures are needed, and these complicate engine design, raising specific weight and cost. It is obvious that in practice the increase of V_{kon} and L will be attained by simultaneous increase of a and P_{ud}, as well as in other ways. It is always necessary to try to develop an engine with the greatest P_{ud} and the least specific weight.

In Fig. 2.2, the curves showing flight altitude H_p* of a long-range missile as functions of the values of a, b, and P_{ud} are shown, no allowance being made for the resistance of the surrounding medium

---

*[$H_п$ = H_p = H_{polet} = H_{flight}.]

and the influence of the acceleration of gravity.

In Table 2.2, there are shown data concerning the tentative dependence of L on a and P_{ud} for the A-4 long-range missile, without allowing for the resistance of the surrounding medium and the acceleration of gravity.

The approximate data given in this table indicate that an increase in a rapidly attains its practical limit, since it would almost be impossible to design a single-stage missile in which 90% of the weight would be due to the fuel, i.e., at a = 90%.

The increase in P_{ud} effectively increases L, in practice, within great limits. In addition, it is, in practice, possible to increase P_{ud} to its upper limits, if not restricted by a fixed energy source.

The range of a liquid-fueled rocket missile of any type is determined to a considerable degree by the velocity V_{kon}.

The development of rocket aerospace equipment is determined by the following magnitudes of flight speed at the end of the powered trajectory*:

a) missiles for hitting bombers and fighters, 1-2 km/sec;

b) missiles for shooting down intercontinental rockets, about 3-7 km/sec;

c) long-range missiles (L = 1000-10,000 km), about 2-8 km/sec;

d) pilotless aircraft (L = 1000-20,000 km), about 1-6 km/sec;

e) rocket vehicles for artificial satellites, about 8 kg/sec [sic];

f) rockets for space flights, about 11.2 kg/sec [sic].

The optimum solution of these requirements depends primarily upon the magnitude of specific thrust attained.

Each of the above-mentioned types of rocket missiles requires different design solutions in the process of development. However, all

---

*Voprosy raketnoy tekhniki, 1958, No. 5, IL.

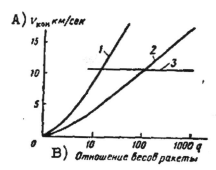

A) $V_{кон}$ км/сек

B) Отношение весов ракеты q

Fig. 2.3. Flight speed of single-stage rocket at end of fuel cutoff, as function of q. 1) $P_{ud} = 300$ kg sec/kg; 2) $P_{ud} = 600$ kg sec/kg; 3) escape velocity. A) V_{kon}, km/sec; B) rocket-weight ratio.

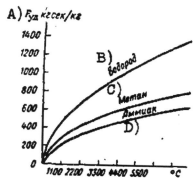

A) $P_{уд}$ кгсек/кг

B) Водород
C) Метан
D) Аммиак

Fig. 2.4. Theoretical specific thrust, dependent on temperature of mass carrier in the reactor of an atomic engine. A) P_{ud}, kgsec/kg; B) hydrogen; C) methane; D) ammonia.

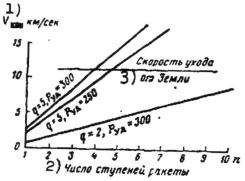

1) $V_{кон}$ км/сек

Скорость ухода
3) от Земли

$q = 5, P_{уд} = 300$
$q = 5, P_{уд} = 250$
$q = 2, P_{уд} = 300$

2) Число ступеней ракеты

Fig. 2.5. Flight speed at end of fuel cutoff attained by multi-stage rockets having various q and specific thrusts. 1) V_{kon}, km/sec; 2) number of rocket stages; 3) escape velocity.

the problems connected with them may be solved in principle by the use of molecular chemical fuels.

In Fig. 2.3 there are shown the roughly calculated values of attainable flight speeds of a single-stage rocket at the end of fuel cutoff, allowing for atmospheric resistance and the effect of gravity, as a function of the weight ratio $q = G_0/G_k$,* where G_0 is the takeoff weight of the missile and the fuel, and G_k is the weight of the missile at the end of fuel cutoff.**

The curves shown in this figure indicate that at $P_{ud} = 300$ kg sec/kg, in order to attain the cosmic (11 km/sec) [escape] velocity altitude a single-stage rocket with a clearly unrealistic weight ratio of $q \geq 150$ is required. At $P_{ud} = 600$ kg sec/kg the development of a rocket with a weight ratio of $q = 9$ is conceivable from a practical standpoint (for example, for a flight to the moon).

The theoretical values of P_{ud} depending on reactor temperature of an atomic engine and the properties of its mass carrier (the energy heat carrier) are shown in Fig. 2.4.

Figure 2.5 gives the calculated approximate values of the attainable flight speeds for multistage rockets at the end of fuel burnout.

The curves in this figure indicate that with a weight ratio of $q = 2$ and with $P_{ud} = 300$ kg sec/kg, the "cosmic" flight velocity can be attained only with 12-13-stage rockets. At $q = 5$ and $P_{ud} = 300$ kg sec/kg, three- or four-stage rockets (with ZhRD) are required for space flights.

The principal task in designing and constructing ZhRD is to achieve the highest possible values of pressure and gas temperature in the combustion chamber, compatible with the required chamber strength. The level of perfection of the chamber nozzle, its cooling method, and other factors substantially affect the magnitude of P_{ud}.

Development of an improved missile design requires new design solutions and more improved methods of production of the missile itself as well as the engine.

---

*[$G_к = G_k = G_{konets} = G_{end}$.]
**Aviation Age, Vol. 28, No. 1, January 1957.

SECTION 8. BASIC PROBLEMS TO BE SOLVED IN PLANNING AND DESIGNING ZhRD

The basic problems which must be solved in planning and designing a liquid-fueled rocket engine for a craft of any designation can generally be broken down as follows.

For a rational choice:

a) type of fuel, system of its supply and atomization;

b) basic parameters of the engine, providing for a solution of the problem posed (pressure in the combustion chamber and in the nozzle outlet section and others);

c) combustion-chamber and engine-nozzle shapes;

d) method (system) of engine cooling;

e) systems for starting and controlling the engine's operation, switching on various valves (electrical, hydraulic, pneumatic) and other elements;

f) designs of the separate elements of the engine;

g) materials used for manufacturing the engine.

For calculations to determine:

a) characteristics of the engine (altitude, throttling, hydraulic, and others);

b) geometrical dimensions of the combustion chamber and nozzle of the engine;

c) parameters and geometric dimensions of the fuel-feed and atomization systems as well as of the engine cooling duct.

The solution of these problems is usually based:

1) on proved theoretical and operational data, and

2) on the results of subsequent tests carried out on the projected and built prototype for purposes of checking and adjusting to achieve economy and reliability of operation.

In determining the basic parameters and dimensions of the engine being designed, we generally make use of the relative data (coefficients) applicable to contemporary engines.

In designing a ZhRD, it is necessary to follow departmental and all-union standards and norms, as well as the technical conditions for manufacture, assembly, and testing.

It is impossible to give general rules for designing a ZhRD, because the development of each separate construction of an engine depends on its use, type of fuel, available data, and other factors. However, the usual methods of choosing the shape of the combustion chamber and nozzle of the engine, the values of its basic parameters, as well as calculations of heat transfer, operational, geometrical, and hydraulic characteristics, systems for fuel feed and injection, are, with some exceptions, suitable for all types of ZhRD.

In designing an engine, we devote great attention to the problems of simplicity and cheapness of construction, convenience of guidance and operation, reliability and economy of operation, and so forth.

The solution of the problem of developing a reliable construction for a ZhRD of great thrust is impeded by the fact that it is attended with constant intensification of the heat processes in the combustion chamber, increase of fuel consumption, and the use of more efficient fuels. An expedient choice of fuel components for a specific engine intended to fulfill a certain function is the first and most obligatory condition for success in developing an engine. However, the greatest difficulties arise in working out the design of the engine.

After completion of the engine design to meet the requirements of the selected fuel type, structural and similar measures are worked out to provide for stable combustion within the combustion chamber of the engine, without fluctuations in pressure and detonation phenomena capable of destroying the engine (because of powerful vibrations and explosions within the engine) during the first seconds of engine operation as it enters the nominal regime, or at any subsequent instant of time.

The following basic stage of engine development is the solution of the problem of cooling the combustion chamber and nozzle. Reliable cooling of a ZhRD always presents great difficulties in developing any engines of prolonged operation, which develop a considerable specific thrust.

The following stage is the development of the engine fuel-feed system and, finally, the last stage involves the development of systems for starting and controlling the engine.

Successful solution of these problems is possible only upon condition of considering the latest scientific achievements in the development of ZhRD and in other associated fields of engineering.

Chapter III
ENGINE OPERATING CYCLES AND EFFICIENCY

The work of a <u>ZhRD</u> consists of changing the chemical energy of liquid fuel into heat energy and then into the kinetic energy of combustion products flowing from the nozzle into the atmosphere, as a result of which a reactive force — the engine's thrust — is developed.

This change of the chemical energy of the fuel in a <u>ZhRD</u> into the other corresponding forms of energy, as in any other engine, is, in practice, accompanied by an unproductive expenditure of part of the energy of the fuel burned. The smaller the magnitude of this loss, the more perfect the engine.

In the present chapter, the thermodynamic processes which form the basis of the change of the chemical energy of the fuel in a <u>ZhRD</u> into the kinetic energy of the gas flow are considered.

SECTION 1. OPERATING CYCLE OF AN IDEAL ENGINE

By the operating cycle of a <u>ZhRD</u> is meant the totality of the thermodynamic processes in the mass carrier in the engine chamber, as a result of which the chemical energy of the fuel is turned into the kinetic energy of the gases issuing from the nozzle into the surrounding medium.

To ascertain and analyze the most important parameters which influence the operating economy of the engine and serve for comparative evaluation of real engines, the concept of the operating cycle of an ideal engine has been introduced into the theory of the ZhRD.

By the operating cycle of an ideal engine, we mean, conventionally, a certain closed and reversible thermodynamic cycle consisting of the simplest thermodynamic processes and constituting a simplified scheme of the combination of the actual processes occurring in real ZhRD.

In accordance with this, the following assumptions are accepted for the operating cycle of an ideal engine:

1) the fuel components are compressed and fed to the combustion chamber without hydraulic resistances in the line and with a negligible loss of energy on these processes, relative to the mechanical work performed by the products of fuel combustion;

2) by means of atomization and mixing of the fuel components fed to the combustion chamber, an absolutely uniform (homogeneous) fuel mixture is formed.

3) the fuel components flow into the combustion chamber at a constant rate;

4) the fuel in the engine's chamber burns at constant pressure and with complete heat liberation (when $\varphi_k = 1*$);

5) the combustion products of the fuel form an ideal gas;

6) the products of combustion expand adiabatically in the nozzle, i.e., without transfer of heat to the surrounding medium, and without complete fuel combustion, recombination, relaxation, and without viscosity friction between the molecules of the gases;

7) the same fields of pressure, temperature, and velocity are to be found in any cross section of the combustion chamber and nozzle throughout their lengths;

_____

$*[\varphi_{\kappa} = \varphi_k = \varphi_{kamera} = \varphi_{chamber}.]$

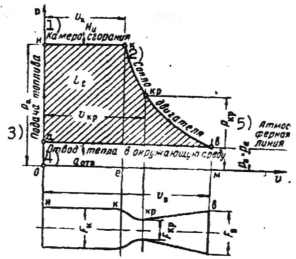

Fig. 3.1. Operating cycle of an ideal ZhRD in pv coordinates. 1) Combustion chamber; 2) engine nozzle; 3) fuel delivery; 4) removal of heat to surrounding medium; 5) atmosphere line.

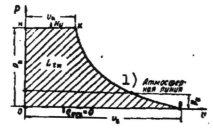

Fig. 3.2. Operating cycle of an ideal ZhRD in space (with complete expansion of gases in the nozzle). 1) Atmosphere line.

8) the movement of the gases in the nozzle outlet is one-dimensional and the gas flow lines are parallel to each other;

9) all the heat is used in the engine with the exception of the heat escaping with the outgoing gas.

With these assumptions, the operating cycle of an ideal engine consists of one isochor, two isobars, and one adiabatic line (Fig. 3.1):

1) the isochor of fuel pressurization and feed into the combustion chamber, characterizing the processes in the fuel-supply system of the engine (line an [ан]);

2) the isobar of fuel combustion in the engine's chamber (line nk [нк];

3) the adiabatic line of the gases' expansion in the engine's nozzle (line kv [кв]);

4) the isobar of the heat transfer from the mass carrier to the surrounding medium, representing the conditional closure of the operating cycle (line va [ва]).

All the basic parameters characterizing the work of one kilogram of gas in the cycle of an ideal engine have values of maximum quantity, and the degree to which the parameters of a real engine approach these determines the perfection of the latter.

The full useful work of one kilogram of gas in the cycle of an ideal engine is expressed in Fig. 3.1 by the area of the shaded diagram, and is expressed analytically thus:

$$L_t = \text{area } \text{онке} + \text{area } \text{екви} - \text{area } \text{оавм} = p_k v_k +$$
$$+ \frac{1}{k-1}(p_k v_k - p_\text{в} v_\text{в}) - p_\text{в} v_\text{в} = \frac{k}{k-1}(p_k v_k - p_\text{в} v_\text{в}) =$$
$$= \frac{k}{k-1}(R_k T_k - R_\text{в} T_\text{в}) = \frac{k}{k-1} R_k (T_k - T_\text{в}) = \frac{k}{k-1} R_k T_k \left(1 - \frac{T_\text{в}}{T_k}\right) =$$
$$= \frac{k}{k-1} R_k T_k \left[1 - \left(\frac{p_\text{в}}{p_k}\right)^{\frac{k-1}{k}}\right] = \frac{v_\text{в.и}^2}{2g} \text{ kg-m/kg.}$$

The full work of one kilogram of gas in the cycle of an ideal engine when operating in space with complete expansion of gas in the nozzle ($p_v = p_a = 0$)* would be

$$L_{t\,\text{п}} = \frac{k}{k-1} R_k T_k = \frac{k}{k-1} p_k v_k = \frac{w_\text{max}^2}{2g} \text{ kg-m/kg,}*** \qquad (3.1)$$

where $w_{v.1}$** is the ideal velocity of gas outflow from the engine's chamber nozzle; w_max is the maximum ideal velocity of gas outflow from the nozzle with complete expansion (up to $p_v = 0$).

The accepted evaluation of the economy of a real engine in relationship to quantity L_t is its operating cycle as accomplished with the same gas pressures in the combustion chamber and in the outlet nozzle, but with fuel energy losses and other deviations from the conditions of the operation of an ideal engine.

The operating cycle of an ideal engine in space (with complete expansion of the gases in the nozzle) is shown in Fig. 3.2.

The degree of perfection for an ideal ZhRD is conventionally evaluated in terms of the thermal efficiency η_t which takes into consideration only the heat losses in the fuel due to the escape of the gases to the surrounding medium, according to the second law of thermodynamics.

SECTION 2. OPERATING CYCLE OF A REAL ENGINE

The operating conditions of a real engine differ considerably from the operating conditions of an ideal engine. In the operation of a real engine one must consider:

1) hydraulic losses in the line during delivery of liquid fuel components to the combustion chamber;

2) the heterogeneity of the velocity and concentration fields of the fuel components in the cross sections throughout the length of the combustion chamber as a result of their imperfect atomization and mixing;

3) the fuel's heat losses in the form of dissociation, incomplete combustion, and heat transfer to the surrounding medium;

4) the mass carrier's pressure difference along the length of the combustion chamber as a consequence of hydraulic losses and flow acceleration during heating up;

5) the complete combustion of the fuel components in the nozzle which were not fully consumed in the chamber because of their imperfect atomization and mixing, and due to their limited chamber stay time.

6) the establishment of a chemical and energy state of equilibrium of the fuel combustion products as they flow and expand in the nozzle;

7) losses as a consequence of the viscosity friction of the gases, i.e., the molecules' friction against each other and the wall of the chamber;

8) the irregularity of the outflow of gases in the supercritical part of the nozzle and the nonparallelism of the flow of gases into the surrounding medium;

9) The underexpansion or overexpansion of the gas in the nozzle as a consequence of various operating regimes and external conditions of the operation of the engine, and other factors.

Because of this the operating cycle of the real engine differs from the cycle of an ideal engine (see Fig. 3.1), and the total work L_1 of one kilogram of gas in it is less than the theoretically conceivable total work L_t of an ideal cycle by the magnitude of the auxiliary losses L_{pot}* in fuel energy, i.e.,

$$L_i = L_t - L_{пот}. \tag{3.2}$$

The work L_1 of the cycle determines the specific thrust of the engine. The greater L_1 is, the greater the engine's P_{ud}.**

The magnitude of auxiliary heat losses of the fuel in a real engine L_{pot} is most affected by:

1) the design of the chamber's burner cup (the method of atomization and mixing of the fuel components, and the type and distribution of the sprayers in the chamber head);

2) the chemical composition of the fuel components and their relationship to each other;

3) the volume of the combustion chamber and its shape;

4) the geometrical characteristics of the nozzle and its profile;

---

*$[L_{пот} = L_{pot} = L_{poterya} = L_{loss}.]$
**$[P_{уд} = P_{specific}.]$

5) the external operating conditions of the engine (operation at land or sea level or at some altitude above the earth), and other factors.

Because of a number of enumerated factors accompanying and complicating the operating cycles in the combustion chamber and nozzle of a real engine, it is practically impossible to precisely define the parameters of the mass carrier even in the characteristic sections of the engine's chamber; therefore, in planning an engine's design, the derivation of a number of assumptions and coefficients simplifying the solution of the problem is often required.

SECTION 3. CLASSIFICATION OF VELOCITIES OF GAS OUTFLOW FROM THE CHAMBER NOZZLE OF A ZhRD

The process of gas outflow from a chamber nozzle of an engine is characterized by the magnitude of the flow's absolute velocity at the nozzle.

To evolve formulas for the efficiency of an engine, it is necessary to understand the physical essence of the conditional concepts of the velocities of gas outflow from an engine chamber nozzle.

In ZhRD theory, the following distinct gas outflow velocities are accepted:

1) maximum ideal w_{max}; 2) ideal $w_{v.1}$; 3) theoretical $w_{v.t}$*; 4) true w_v*; 5) effective w_{ef}*; 6) critical w_{kr}.*

Maximum ideal velocity is that exhaust velocity which a gas would have in an ideal engine's chamber nozzle outlet section with complete change of the available chemical energy of one kilogram of fuel into

___

*[$w_{в.т} = w_{v.t} = w_{vykhod.teoreticheskaya} = w_{outlet.theoretical}$;
$w_{в} = w_v = w_{vykhod} = w_{outlet}$;
$w_{эф} = w_{ef} = w_{effektivnaya} = w_{effective}$;
$w_{кр} = w_{kr} = w_{kriticheskaya} = w_{critical}$.]

the kinetic energy of the gas at an infinitely high level of expansion, i.e.,

$$w_{max} = \sqrt{2gH_s} = 91{,}53\sqrt{H_s} = \sqrt{2g\frac{k}{k-1}R_\kappa T_\kappa} \text{ m/sec},\qquad(3.3)$$

where H_u kcal/kg is the lowest calorific value of the fuel.

The magnitude of the ideal exhaust velocity is characteristic of the fuel's properties.

Ideal velocity is the exhaust velocity which a gas would have in an ideal engine's chamber nozzle outlet section, assuming adiabatic change of condition and allowing for the counter-pressure of the atmospheric air; i.e.,

$$w_{s.u} = \sqrt{2gH_u\tau_u} = 91{,}53\sqrt{H_u\tau_u} = \sqrt{2g\frac{k}{k-1}R_\kappa T_\kappa\left[1-\left(\frac{p_s}{p_\kappa}\right)^{\frac{k-1}{k}}\right]} =$$
$$= \sqrt{2g\frac{k}{k-1}R_\kappa T_\kappa \tau_u} = w_{max}\sqrt{\tau_u} \text{ m/sec},\qquad(3.4)$$

where η_t is the thermal efficiency of the engine.

Theoretical velocity is the exhaust velocity of the chemically active gases in the chamber nozzle outlet of a real engine, theoretically calculated with regard to only one fuel heat loss in the combustion chamber in the form of dissociation of gases, and with the assumption of complete chemical- and energy-equilibrium isentropic change in the state of the products of combustion in the chamber nozzle, i.e.,

$$w_{s.\tau} = \sqrt{2g\frac{k}{k-1}R_{\kappa.\tau}T_{\kappa.\tau}\left[1-\left(\frac{p_s}{p_\kappa}\right)^{\frac{k-1}{k}}\right]} = 91{,}53\sqrt{I_\kappa - I_s} \text{ m/sec},$$

where k is the mean value of the exponent of the level of the isentrope equation; I_k and I_v* are the energy content of the gases in the combustion chamber and the chamber nozzle outlet section, respectively, of a real engine in kcal/kg.

----

*[$I_\kappa = I_{chamber}$; $I_s = I_{outlet}$.]

True velocity is the mean velocity of gas outflow in the chamber nozzle outlet of a real engine, usually defined with the assumption of polytropic expansion of the gases in the nozzle:

$$w_s = \sqrt{2gH_s 427\eta_i} = 91{,}53\sqrt{H_s \eta_i} =$$

$$= \sqrt{2g\frac{n}{n-1}R_{\kappa}T_{\kappa}\left[1-\left(\frac{p_a}{p_{\kappa}}\right)^{\frac{n-1}{n}}\right]} \; m/sec, \qquad (3.5)$$

where η_i is the internal efficiency of the engine.

In deriving equations for determining the velocity of gas outflow from an engine chamber nozzle, the following are disregarded:

1) the velocity at which fuel components are injected into the combustion chamber, as a consequence of their negligibly small magnitudes relative to gas exhaust velocity from the nozzle;

2) the heat resistance of the engine's combustion chamber (the drop in pressure along the length of the chamber because of the combustion products' heating up and attaining a movement up to the velocity w_k*), which, if necessary, is calculated by a special coefficient (see Section 13, Chapter VI [sic]);

3) energy loss of the gas flow as a consequence of radially composed velocity at the exit from the expanding part of the chamber nozzle, which is usually also computed by a special coefficient.

The velocity w_v is the summary of the characteristics of an engine's operating efficiency.

In existing ZhRD $w_v \approx$ 2000-2500 m/sec. For the most efficient chemical fuels' products of combustion, the possible exhaust velocity amounts to about 4.5 km/sec.

Obtaining the maximum velocity of exit gases from the chamber nozzle is one of the basic tasks of the designer of an engine.

---

*[w_k = $w_{chamber}$.]

The greater the velocity w_v, the greater the specific thrust of the engine proportional to it, since $P_{ud} = w_v/g \approx 0.1 w_v$ (when $p_v = p_a$).*

<u>Effective</u> velocity is the conditional velocity of the gases behind the chamber nozzle of a real engine, computed in accordance with the general equation for the thrust of an engine working on the corresponding operating regime and flight altitude:

$$P_{\text{н}} = \frac{G_s}{g} w_{\text{н}} + F_{\text{н}}(p_{\text{н}} - p_{\text{н}}) = \frac{G_s}{g} w_{\text{эф}}, \qquad ** \qquad (3.6)$$

i.e.,

$$w_{\text{эф}} = \frac{gP_{\text{н}}}{G_s} = w_{\text{н}} + \frac{gF_{\text{н}}}{G_s}(p_{\text{н}} - p_{\text{н}}) = gP_{\text{уд.н}} \text{ m/sec.} \quad *** \qquad (3.7)$$

When $p_v = p_a$ we have $w_{ef.n} = w_{ef} = w_v$. ****

One may distinguish the following effective exhaust velocities:

1) at ground or sea level w_{ef_0};

2) in flight at a specified altitude $w_{ef\ H}$;

3) during flight in space $w_{ef.p}$. *****

<u>Critical</u> velocity is the velocity of gas outflow in the outlet of the engine chamber nozzle:

$$w_{\text{кр}} = \sqrt{2g \frac{n}{n+1} R_{\text{к}} T_{\text{к}}} = \sqrt{ngRT_{\text{кр}}} \text{ m/sec.} \qquad (3.8)$$

Increase of velocity w_v is, in principle, possible by increasing parameters H_u, R_k, $\underline{n}$, p_k/p_v, T_k, and η_1. Some of these parameters are, in practice, extremely limited.

Within the widest limits, one may increase w_v by using fuel of high calorific values in the <u>ZhRD</u>, as well as fuels whose combustion

---

*[$p_a = p_a = p_{\text{atmosfera}} = p_{\text{atmosphere}}$.]

**[$P_{\text{н}} = P_n = P_{\text{neogranichennaya}} = P_{\text{absolute}}$.]

***[$gP_{\text{уд.н}} = gP_{ud.n} = gP_{\text{udel'naya.neogranichennaya}} = gP_{\text{specific.absolute}}$.]

****[$w_{\text{эф.н}} = w_{ef.n} =$

*****[$w_{\text{эф.п}} = w_{ef.p} = w_{\text{effektivnaya.pustota}} = w_{\text{effective.space}}$.]

products, other things being equal, have the greatest value of R_k.

The use of fuels of high calorific value to increase exhaust velocity, other things being equal, usually raises combustion temperature and therefore requires mor intensive cooling of the engine chamber, more intensive removal of heat from the heated inner liner of the chamber, and the use of more expensive heat-resistant metals for manufacturing the chamber.

When the chamber liner has a high heat liberation rate, an engine cooling system which is more complicated in construction, relatively expensive, and less economical may be required.

Increase of gas exhaust velocity by raising pressure in the combustion chamber above a determined limit, different for every type of engine, is limited in practice because of the increase of engine weight and the difficulties of cooling the chamber.

Increasing w_v by decreasing gas pressure p_v at the outlet from the engine nozzle is also, in practice, limited by the tactical purpose of the weapon, engine operating conditions relative to the surrounding medium, and other factors.

In existing engines, pressure difference in the nozzle amounts to $p_k/p_v \approx 16-60$ and more. Further increase of the initial gas pressure in the combustion chamber leads to a comparatively small increase of exhaust velocity but, in return, to a more considerable increase of the specific weight of the engine. Therefore, one may scarcely expect that from the economic point of view a pressure above 60-100 atmospheres absolute would be used in the combustion chamber. By decrease of the gas pressure in the engine nozzle outlet, the blastoff thrust of the engine is lowered and nozzle weight is increased, so that decreasing pressure is, in each separate case, held within fixed limits.

SECTION 4. ENGINE EFFICIENCIES

The accepted definition of efficiency is the degree of perfection, in a <u>ZhRD</u>, of the change of the chemical energy of the fuel into the kinetic energy of the combustion products at the outlet from the chamber nozzle: thermal efficiency η_t; internal efficiency η_1; relative efficiency η_0*; mechanical efficiency η_m, and effective efficiency η_e. Besides this, there are additional efficiencies which have been accepted to define the engine's work during the missile's flight: thrust efficiency η_p and total efficiency η_{ob}.*

<u>Thermal</u> efficiency η_t of the engine shows what part of the available chemical energy of the fuel fed to the combustion chamber would be changed to the kinetic energy of the gases at the outlet from the chamber nozzle in case of an ideal <u>ZhRD</u>, i.e.,

$$\eta_t = \frac{L_t}{H_u 427} = \frac{w_{s.н}^2/2g}{w_{max}^2/2g} = \left(\frac{w_{s.н}}{w_{max}}\right)^2 =$$

i.e.,

$$= \left(\frac{\sqrt{2g\frac{k}{k-1}R_к T_к\left[1-\left(\frac{p_ь}{p_к}\right)^{\frac{k-1}{k}}\right]}}{\sqrt{2g\frac{k}{k-1}R_к T_к}}\right)^2 = 1-\left(\frac{p_ь}{p_к}\right)^{\frac{k-1}{k}}.$$

$$\eta_t = \left(\frac{w_{s.н}}{w_{max}}\right)^2 = 1-\left(\frac{p_к}{p_к}\right)^{\frac{k-1}{k}}.$$

(3.9)

Consequently, when determining an engine's thermal efficiency, only the heat loss of the working fluid as stated in the second law of thermodynamics is considered.

The value of η_t depends on the expansion level p_k/p_v of the gases in the nozzle and the adiabatic exponent <u>k</u> (depending on the type of fuel).

---

*$[\eta_0 = \eta_o = \eta_{otnositel'nyy} = \eta_{relative};$
$\eta_{об} = \eta_{ob} = \eta_{obshchiy} = \eta_{total}.]$

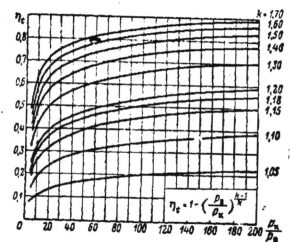

Fig. 3.3. Thermal efficiency η_t of an engine, as a function of p_k and k.

In practice, different fuels at the same conditions have different speeds of combustion and recombination, which are differentiated by the composition of the gases, their specific heat, and, consequently, by the magnitude of k.

In Table 3.1 and in Fig. 3.3 the values of η_t when p_v = 1 atm-abs and at various magnitudes of p_k and k are shown. The data in the table and the curves of Fig. 3.3 show that when p_k is increased, the value of η_t, at first, grows rapidly (in direct proportion to k), and then slower, and at about 200 atm-abs becomes almost constant.

Consequently, to obtain the greatest value of η_t it is necessary for the engine to operate at the highest possible p_k, and also using a fuel whose combustion products have the greatest possible specific heat ratio (k = c_p/c_v).*

With a rise in p_k the pressure difference p_k/p_v in the nozzle is increased and the operating cycle in the engine is intensified, basically by decreasing the dissociation of the gases in the combustion chamber. In some cases an increase of p_k to 100 atm-abs may be justified for decreasing the dimensions and weight of the engine chamber.

*[The specific heat ratio k, not to be confused with the adiabatic exponent k just mentioned. The specific heat ratio k is a Russian lower-case italic letter.]

TABLE 3.1

Values of η_t Depending on p_k and $\underline{k}$ when $p_v =$
$= 1$ Atmosphere Absolute

k \ $p_v/p_к$	1/5	1/10	1/20	1/50	1/100	1/200
1,0	0	0	0	0	0	0
1,05	0,074	0,104	0,133	0,150	0,197	0,223
1,10	0,136	0,189	0,238	0,299	0,342	0,382
1,15	0,189	0,259	0,323	0,400	0,451	0,499
1,18	0,218	0,196	0,367	0,449	0,505	0,554
1,20	0,232	0,319	0,393	0,479	0,536	0,586
1,30	0,310	0,412	0,449	0,595	0,654	0,706
1,40	0,369	0,482	0,575	0,673	0,732	0,780
1,50	0,415	0,536	0,632	0,729	0,785	0,829
1,60	0,453	0,578	0,675	0,769	0,822	0,863
1,70	0,482	0,613	0,709	0,800	0,850	0,887

If the engine operates with underexpansion of the gases in the nozzle $(p_v > p_a)$, in this case the thermal efficiency is expressed by the formula

$$\eta_{\text{нед}} = \varphi_{\text{нед}}\eta_t \quad * \tag{3.10}$$

where φ_{ned} is the factor of completeness of gas expansion in the nozzle relative to atmospheric pressure:

$$\varphi_{\text{нед}} = \frac{1-(p_v/p_к)^{\frac{k-1}{k}}}{1-(p_a/p_к)^{\frac{k-1}{k}}};$$

here p_v is the pressure in the outlet of the engine nozzle with underexpansion of the gases $(p_v > p_a)$; and p_a is the pressure of the atmospheric air.

Existing engines have a $\eta_t \approx 0.40-0.70$.

<u>The internal</u> efficiency of an engine shows what part of the available chemical energy of the fuel fed to the combustion chamber of a

---

$*[\varphi_{\text{нед}} = \varphi_{ned} = \varphi_{nedorasshireniye} = \varphi_{underexpansion};$
$\eta_{\text{нед}} = \eta_{underexpansion}.]$

real engine is turned into kinetic energy of the gases in the outlet from the nozzle, i.e.,

$$\eta_i = \frac{L_i}{H_u 427} = \frac{w_v^2/2g}{w_{max}^2/2g} = \left(\frac{w_v}{w_{max}}\right)^2.$$ (3.11)

When $p_v \neq p_a$, the value of w_v must be replaced by the value of w_{ef}. Since when $p_v = p_a$ the absolute thrust of the engine chamber is

$P = G_s/g(w_v)$, from which $w_v = gP/G_s$ m/sec,

then

$$\eta_i = \frac{w_v^2/2g}{H_u 427} = \frac{\left(\frac{gP}{G_s}\right)^2/2g}{H_u 427} = \frac{P^2}{87,05 H_u G_s^2}.$$

Since $P_{ud} = P/G_s$, then

$$\eta_i = \frac{P_{yd}^2}{87,05 H_u} = 0,0115 \frac{P_{yd}}{H_u}.$$ (3.12)

If P and G_s have been measured during the engine's trials on a test stand, it is possible to compute η_1 and w_v extremely precisely by the latter equation.

Existing engines have an $\eta_1 \approx 0.30$-0.50.

Relative efficiency of an engine shows the degree of deviation of the operating cycle in the chamber of a real engine from the operating cycle in the chamber of an ideal engine, i.e.,

$$\eta_o = \frac{L_i}{L_e} = \frac{w_v^2/2g}{H_u 427 \tau_{il}} = \frac{\tau_{il}}{\tau_{il}}.$$ (3.13)

This efficiency considers the losses of fuel energy in a real engine which are conditional upon the physical incompleteness of fuel combustion as a result of imperfect mixing of its components and inadequate speed of combustion and recombination of gases during expansion in the nozzle, as well as losses caused by the energetic instability of outflow of gases from the nozzle and the nonadiabatic character of the process (heat transfer to the surrounding medium and heat transfer

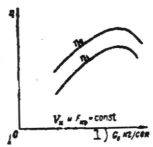

Fig. 3.4. Factors η_1
and η_0 dependent on
G_s of an engine,
with unchanging val-
ues of V_k and F_{kr}.
1) G_s, kg/sec.

to the liner as a result of friction and
radiation).

Existing engines (the "Vanguard" missile
and others) have a $\eta_0 \approx 0.65-0.90$.

By raising p_k, the values of η_1 and η_0
are increased by increasing the fuel's heat
liberation coefficient φ_k. Increase of p_k by
increasing G_s without changing the dimensions
of the combustion chamber and nozzle of the
engine leads at first to an increase of η_1 and

η_0 because the operating regime is close to the optimum regime in regard to flow rate of the fuel in a given combustion chamber's volume,
and then to their decrease as a consequence of the discrepancy of the
volume of the combustion chamber V_k to the ever-increasing fuel flow
rate G_s per second and the decrease of φ_k by this means (Fig. 3.4).

Mechanical efficiency of an engine considers the decrease of the
internal efficiency of the engine caused by possible additional flow
rates in serving the fuel-feeder system and in forming the protective
curtain around the surface of the burner liner.

When $p_s = p_a$,* the specific thrust may be expressed as follows:

1) combustion chamber:

$$P_{y1.s} = \frac{P_k}{G_s} = \frac{w_s}{g} = \frac{91{,}53\sqrt{H_s \eta_i}}{g} = 9{,}33\sqrt{H_s \eta_i}\ \frac{\text{kg thrust}}{\text{kg fuel/sec}};$$

2) engine as a whole:

$$P_{y1z} = \frac{P_z}{G_z} = 9{,}33\sqrt{H_s \eta_i \eta_m}\ \frac{\text{kg thrust}}{\text{kg fuel/sec}},$$

where G_Σ is the total flow rate in the engine in kg/sec.

The absolute thrust of an engine with a <u>TNA</u> may be greater than

*[$p_c = p_s = p_{sistema} = p_{system}.$]

78

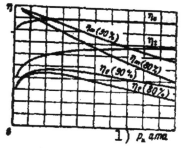

Fig. 3.5. Engine efficiency η_t, η_1, η_0, η_e, and η_m depending on p_k. 1) Atmosphere absolute.

the thrust of an engine without a <u>TNA</u>, if the gases which have worked in the turbine, flowing out into the surrounding medium, create an additional thrust $\Delta P_{\underline{TNA}}$ or are burned in the basic chamber or an auxiliary engine chamber.

Having divided the expression mentioned above for $P_{ud.k}$ into the expression for $P_{ud\Sigma}$, we will obtain a formula for computation of the mechanical efficiency of an engine, i.e.,

$$\eta_m = \left(\frac{P_{yд\,э}}{P_{yд.\kappa}}\right)^2 = \left(\frac{P_э}{P_\kappa}\right)^2\left(\frac{G_s}{G_s + G'_s}\right)^2 = \left(\frac{P_э}{P_\kappa}\right)^2\left(\frac{1}{1 + G_s/G'_s}\right)^2 = \left(\frac{P_э}{P_\kappa}\right)^2\left(\frac{1}{1 + \zeta_{rr}}\right)^2,\qquad(3.14)$$

whence

$$P_{yд\,э} = P_{yд.\kappa}\sqrt{\eta_m} = P_{yд.\kappa}\frac{P_э}{P_\kappa}\frac{1}{1 + \zeta_{rr}}\ \frac{\text{kg thrust}}{\text{kg fuel/sec}},$$

or when $P_\Sigma = P_k$, we will obtain

$$P_{yд\,э} = P_{yд.\kappa}\frac{1}{1 + \zeta_{rr}}\ \frac{\text{kg thrust}}{\text{kg fuel/sec}},$$

where $\zeta_{gg} = G'_s/G_s \approx f(P_k, p_p)$* is the relative fuel consumption in serving the fuel-feed system (gas generator) of the engine.

In existing engines with turbopump units we have:

1) in missile engines $\zeta_{gg} \approx 0.015-0.030$;

2) in aircraft engines $\zeta_{gg} \approx 0.050-0.060$.

For existing missile engines feeding the turbine of the turbopump unit with gas vapor from 80% hydrogen peroxide or fuel combustion products removed from the engine combustion chamber (the "Atlas" mis-

*[$\zeta_{rr} = \zeta_{gg} = \zeta_{gazogenerator} = \zeta_{gas\ generator}$;

$p_\pi = p_p = p_{podacha} = p_{feed}$.]

sile and others) $\eta_m \approx 0.97-0.98$.

By raising p_k, the value of η_m is decreased almost in accordance with the linear law (Fig. 3.5). The more effective an engine's fuel-feed system is, the greater its η_m.

In engines with a closed system for cooling and feeding the turbine $\eta_m \approx 1$, since the expenditure of energy in starting the circulation pump of this system is negligibly small. For engines supplied by means of a VAD, whose $G_g = 0$, nominally $\eta_m = 1$. However, in comparing the effectiveness of operation of these engines with a ZhRD having a turbopump unit, one should not lose sight of the preliminary expenditure of energy in obtaining the compressed gas for the fuel-feed system.

The effective efficiency of an engine shows what part of the available chemical energy of the total flow rate in a real engine is turned into the kinetic energy of gases as a result of which thrust force is developed; i.e.,

$$\eta_e = \eta_i \eta_m = \eta_i \eta_0 \eta_m. \tag{3.15}$$

This efficiency may be expressed as follows:

$$\eta_e = \frac{L_i}{\left(H_s + H'_s \dfrac{G'_s}{G_s}\right) 427} = \frac{w_e^2 / 2g}{(H_s + H'_s \zeta_{rr}) 427}. \tag{3.15'}$$

If the same fuel which is expended in the combustion chamber of the engine is used to start the turbopump unit, $H_u = H'_u$ and

$$\eta_e = \frac{w_e^2/2g}{(1 + \zeta_{rr}) H_s 427} = \frac{\eta_i}{1 + \zeta_{rr}}.$$

from whence

$$w_e = \sqrt{2g H_s 427 (1 + \zeta_{rr}) \eta_e} \quad \text{m/sec.}$$

Some engines have a G'_g equal to zero, and, consequently $\eta_e = \eta_1$. Existing engines have a $\eta_e \approx 0.25-0.45$ and over.

In a number of ZhRD now manufactured, 80% hydrogen peroxide is sometimes used to start the TNA.

For an engine in flight, the effective efficiency must be determined by consideration of the kinetic energy of the basic and auxiliary fuels which are moving with the weapon at a velocity V in m/sec, using the formula

$$\eta_{eV} = \frac{w_{э\phi}^2/2g + V^2/2g}{(H_u + H_u'\zeta_{rr})\,427 + \dfrac{V^2}{2g}(1 + \zeta_{rr})} =$$

$$= \frac{w_{э\phi}^2 + V^2}{2g\,(H_u + H_u'\zeta_{rr})\,427 + V^2(1 + \zeta_{rr})}. \tag{3.16}$$

If the TNA operates on the basic fuel, $H_u = H'_u$, and therefore

$$\eta_{eV} = \frac{w_{э\phi}^2 + V^2}{2g\left(H_u 427 + \dfrac{V^2}{2g}\right)(1 + \zeta_{rr})}. \tag{3.16'}$$

With a heat value of the fuel of the order of H_u = 2000 kcal/kg and a flight velocity V_{kon} = 3000 m/sec* (M ≈ 10) the kinetic energy of the fuel amounts to about 50% of its heating value. One may not disregard such a quantity of energy.

<u>The thrust</u> efficiency of an engine shows what part of the available kinetic energy of the gases in the chamber nozzle of a real engine is used during the flight of a weapon, i.e.,

$$\eta_p = \frac{L_{kin.исп}}{L_{kin.pac\pi}} = \frac{L_{kin.исп}}{L_{kin.исп} + L_{kin.неисп}} = \frac{P_{уд} V}{P_{уд} V + \dfrac{(w_{э\phi} - V)^2}{2g}} =$$

$$= \frac{(w_{э\phi}/g)\,V}{\dfrac{w_{э\phi}}{g} V + \dfrac{(w_{э\phi} - V)^2}{2g}} = \frac{2 w_{э\phi} V}{w_{э\phi}^2 + V^2} = \frac{2(V/w_{э\phi})}{1 + (V/w_{э\phi})^2}. \quad ** \tag{3.17}$$

---

*[$V_{КОН}$ = V_{kon} = V_{konets} = V_{end}.]

**[$L_{кин.исп}$ = $L_{kin.isp}$ = $L_{kineticheskaya.ispolzovat'sya}$ =
= $L_{kinetic.used}$;

$L_{кин.pacп}$ = $L_{kin.rasp}$ = $L_{kineticheskaya.raspolagaemaya}$ =
= $L_{kinetic.available}$;

$L_{кин.неисп}$ = $L_{kin.neisp}$ = $L_{kineticheskaya.neispolzovat'sya}$ =
= $L_{kinetic.not\ used}$.]

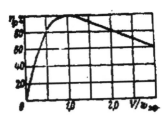

Fig. 3.6. Thrust efficiency of an engine η_p depending on the relationship of velocities V/w_{ef}.

When $V/w_{ef} = 0$, $\eta_p = 0$; when $w_{ef} = V$, $\eta_p = 1$, i.e., it attains the maximum when $w_{ef} = 0$ relative to the earth, and when $V/w_{ef} > 1$, it is decreased (Fig. 3.6).

The A-4 long-range missile, with a flight velocity at the end of the boost phase $V_{kon} = 1525$ m/sec, and a velocity $w_{ef} = 2135$ m/sec had an $\eta_p = 0.94$.

The total efficiency of an engine shows what part of the available energy of the total flow rate in a real liquid-propellant rocket engine is usefully utilized during the weapon's flight, i.e.,

$$\eta_{o6m} = \eta_{oV}\eta_{p}. \text{*} \tag{3.18}$$

The magnitude of the specific thrust of the engine is used for evaluation and comparison between the work of different liquid-fueled rocket engines.

In the theoretical calculation of efficiency and specific thrust of an engine, it is necessary to know the precise values of the heats of reaction and phase transitions, specific heat, and the constant of equilibrium of the combustion products of the given fuel.

The values of all the above-mentioned efficiencies of an engine depend, to a considerable degree, on the type of fuel used. It follows that one must keep in mind that for a ZhRD not the effective efficiency η_e, but the specific thrust, determined by the effective outflow velocity, is distinctive, since cases are possible when by dilution of the fuel mixture with water and lowering its heating value, it is possible to raise even η_t or η_o, and, consequently, even η_e, at the same time lowering P_{ud}. Therefore the best engine is not the one which

---

*[η_{o6m} = η_{obshch} = $\eta_{obshchiy}$ = η_{total}.]

is more economical in using the available heat, but the one which develops the greatest specific thrust.

For this reason, efficiency values may be used only for comparative evaluation of a ZhRD, operating on the same fuel and at identical combustion conditions or in the operation of a given engine with various flight speeds.

Example 1. An engine develops a thrust P = 908 kg at a flow rate G_g = 4.86 kg/sec and H_u = 1665 kcal/kg. *

Flight speed of missile V = 1285 m/sec.

Determine w_{ef}, L_{kin}, η_i, η_p, η_{obshch}, and P_{ud}.

Solution.

1. Effective velocity of gas outflow

$$w_{ef} = \frac{gP}{G_g} = \frac{9.81 \cdot 908}{4.86} = 1840 \text{ m/sec.}$$

2. Kinetic energy of 1 kg of gases behind the engine nozzle

$$L_{kin} = \frac{w_{ef}^2}{2g} = \frac{1840^2}{2 \cdot 9.81} = 17.2 \cdot 10^4 \text{ kgm/kg.}$$

3. Engine efficiency:

a) internal

$$\eta_i = \frac{w_{ef}^2/2g}{H_u 427} = \frac{1840^2/2,9.81}{1665 \cdot 427} = \frac{7.2 \cdot 10^5}{1665 \cdot 427} = 0.242 = 24.2\%;$$

b) thrust

$$\eta_p = \frac{2\frac{v}{w_{ef}}}{1 + \left(\frac{v}{w_{ef}}\right)^2} = \frac{2\frac{1285}{1840}}{1 + \left(\frac{1285}{1840}\right)^2} = 0.375 = 37.5\%;$$

c) total

$$\eta_{obshch} = \frac{PV}{G_g H_u 427 + \frac{V^2}{2g} G_g} = \frac{w_{ef} V}{g H_u 427 + V^2/2} = \frac{2(V/w_{ef})}{\frac{1}{\eta_i} + (V/w_{ef})^2} =$$

$$= \frac{2(1285/1840)}{\frac{1}{0.242} + \left(\frac{1285}{1840}\right)^2} = 0.0936 = 9.36\%.$$

*D. Satton [probably George Sutton], Rocket Engines, IL, 1950.

4. Specific thrust

$$P_{y\text{л}} = \frac{w_{\text{э}\phi}}{g} = \frac{1840}{9.81} = 187 \ \frac{\text{kg thrust}}{\text{kg fuel/sec}}.$$

5. Specific fuel consumption

$$C_{ud} = G_s/P = 4.86/908 = 0.00535 \ \frac{\text{kg fuel/sec}}{\text{kg thrust}} =$$

$$= 5.35 \ \frac{\text{kg fuel/sec}}{\text{tons of thrust}}.$$

SECTION 5. CHANGE OF WORKING FLUID'S BASIC PARAMETERS ALONG THE CHAMBER LENGTH OF A ZhRD

In the operating cycle in a ZhRD, the state of the mass carrier (fuel and combustion products) is changed during its movement along the combustion chamber and nozzle. By this means, the change of the mass carrier's parameters (pressure, temperature, velocity, and others) along the length of the engine chamber is conditional upon the character of the operating cycle occurring in the chamber.

In the majority of existing ZhRD fuel components are fed from the tanks to the combustion chamber under a pressure $p_p \approx 20\text{-}60$ atm abs or above (Fig. 3.7).

The pressure of the fuel components in the cooling jacket and in the injectors in the engine chamber head is decreased to approximately 10-18 atm abs as a consequence of the increase of the velocity of their outflow and the hydraulic resistances in the lines and in the burner cup.

Directly along the length of the combustion chamber, gas-flow pressure is slowly decreased as the result of losses by friction against the surface of the liner and the increase of the velocity of movement under the influence of the inflow of the compressed fuel's heat (and sometimes also by the change in the cross section of the combustion chamber along its length). Decrease of gas pressure

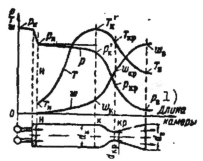

Fig. 3.7. Picture of change of basic parameters of mass carrier (working fluid) along the length of a real engine's chamber. 1) Length of chamber.

along the length of an ordinary cylindrical combustion chamber is insignificant.

Decrease of gas pressure at the end of a cylindrical combustion chamber usually does not exceed 0.5-1% of the pressure at the chamber head. Therefore, in calculating the pressure difference of the gases in the combustion chamber one may accept it as constant, and consider the combustion chamber as an isobaric chamber, if $f_k = F_k/F_{kr} \geq 5-6$.

In the engine nozzle, gas-flow pressure is considerably decreased, since here the geometrical influence on the flow takes place, and at the outlet from the chamber to the atmosphere it ordinarily amounts to 0.6-1.2 atm abs.

Pressure at a cross section of the chamber nozzle is established by considering the external conditions of the engine's use, and therefore must be different for engines of different designations.

The temperature of the fuel components fed from the tanks to the engine combustion chamber (liquid under normal physical conditions) is practically equal to the temperature of the surrounding medium.

If a fuel component is used for cooling the combustion chamber and nozzle of the engine, by this means it is ordinarily heated up by 50-100°C or more, but not higher than its boiling temperature at the given local pressure in the cooling duct.

The fuel components injected into the combustion chamber are, at first, heated up by means of the combustion heat of the previous portions of the fuel, are vaporized, and then by means of mixing with each other are ignited and burn.

The temperature of the gases in the zone where the fuel is prepared for combustion remains almost constant, increases sharply in the combustion zone, and decreases during the gases' expansion in the nozzle. Temperature at the combustion chamber head ordinarily amounts to 600-900°C. In the presence of prechamber vaporization of the fuel components, this temperature is higher.

Temperature of the gases at the end of the combustion chamber (before the nozzle) amounts to 2800-3300°C, and in the outlet section of the nozzle, 1000-2000°C. These temperatures considerably exceed the melting temperature of the materials of the chamber's inner liner, which creates the necessity of cooling it reliably.

Injection velocity w_{vpr}* of the fuel components into the combustion chamber depends chiefly on the feed pressure, design of the vaporizing unit, pressure in the combustion chamber, and other factors, and usually varies from 20 to 40 m/sec.

The velocity of the gases grows slowly along the length of the combustion chamber and is sharply increased in the engine nozzle.

In ZhRD combustion chambers, $w_k \approx$ 50-200 m/sec, and $w_v \approx$ 2000 to 2500 m/sec.

As a consequence of the considerable turbulence of the gas flow and the extremely uneven distribution of the gas-flow velocities in all sections of the combustion chamber and nozzle, one may speak only of the average parameters of the working fluid, somewhat different from their true values.

One must also keep in mind that the computed values of the parameters of the mass carrier in the characteristic sections of the combustion chamber and nozzle of the engine as mentioned above are approx-

---

*[$w_{впр}$ = w_{vpr} = w_{vprysk} = $w_{injection}$.]

imate; in each individual case, it is necessary to determine them either by calculations or experimentally (the latter method is suitable for existing engines).

SECTION 6. EQUATIONS CHARACTERIZING THE QUANTITATIVE RELATIONSHIPS IN THE CHANGE OF GAS-FLOW PARAMETERS ALONG THE LENGTH OF THE ZhRD NOZZLE

In planning and designing a ZhRD, one must know the quantitative relationships in the change of gas-flow parameters (pressure, temperature, velocity, etc.) along the length of the combustion chamber and nozzle.

The connections between the temperatures of the gas flow in various sections of the engine chamber, on the one hand, and velocities, pressures, and densities of this flow, on the other hand, may be expressed by the corresponding equations of gas dynamics. In using these equations, we usually make the assumption that temperatures, velocities, and pressures of the gases in the sections of the flow channel being studied are the same. These assumptions are actually valid only for the nozzle portion of the engine chamber, and are completely inapplicable to the operating conditions of the combustion chamber, where the characteristics of the working fluid are extremely changeable, and do not lend themselves to precise quantitative analysis.

The parameters of the gas flow in the chamber of a ZhRD are changed as a consequence of:

1) geometrical influence on the flow, i.e., changes in the cross section of the channel for the gas flow;

2) heat influence on the flow, i.e., transfer of heat from the gas flow to vaporization of fuel components, to the surrounding medium, and as a result of the dissociation of fuel combustion products, as well as the feeding of heat to the gas flow as the result of fuel com-

bustion and recombination of the products of gas dissociation;

3) *flow-rate influence* on the stream, i.e., increase of the quantity of gas per second as the result of vaporization and combustion of fuel components (the presence of the smallest liquid particles of the fuel in the gas flow may be disregarded because of their relatively trifling amount), and

4) *chemical influence* on the flow, i.e., change of the number of moles per unit mass in the gas stream resulting from the chemical reactions occurring (fuel combustion, dissociation, and recombination of the gas molecules).

The intensity of the above-enumerated influences on the gas flow in a ZhRD is diverse and its quantitative calculation is extremely difficult.

The velocity of the gas flow in the engine is increased basically because of geometrical influence. In the combustion chamber the gas velocity is increased chiefly by means of heat and chemical influences. However, the latter lead to a considerable increase of the velocity of the gas flow only in a high-speed engine combustion chamber, where the relative area $f_k < 3$. The requirement for such a high-speed combustion chamber practically never arises.

Flow-rate influence on the gas stream in a ZhRD nozzle is even less significant.

Since, in the combustion chambers of conventional engines, $f_k > 4$, and, therefore, the velocity of the gas flow at the outlet to the nozzle is comparatively small, in calculations of a ZhRD its value may be accepted as equal to zero. This assumption permits the determination of the parameters of the gas flow at the outlet from the combustion chamber to the nozzle only on the basis of the data of the thermodynamic calculations of the engine. In those cases when this gas velocity cannot be disregarded, its value may be determined after the thermodynamic calculation of the fuel combustion process.

When designing a ZhRD, one ordinarily computes the quantitative relationships of the geometrical influence on the gas flow in the nozzle, the effect of the other influences on the gas being thus calculated by using the polytropic exponent n or by other parameters on the basis of the principles of technical thermodynamics and gasdynamics.

The quantitative relationships of geometrical influence on the gas flow may be most easily obtained in the form of the dependences of the separate gas parameters on the velocity regime, i.e., on the Mach number (M = w/a, where w is the velocity of the stream and a the local speed of sound).

The velocity regime of an adiabatic gas-stream current in the geometrical nozzle of an engine may be expressed as follows:

a) at the entrance to the chamber nozzle

$$M_\kappa = \frac{w_\kappa}{a_\kappa} = \frac{w_\kappa}{\sqrt{kg R_\kappa T_\kappa}} = \frac{w_\kappa}{\sqrt{k \frac{p_\kappa}{\rho_\kappa}}};$$ (3.19)

b) in any nozzle cross section

$$M = \frac{w}{a} = \frac{w}{\sqrt{kg R T}} = \frac{w}{\sqrt{k \frac{p}{\rho}}},$$ (3.19')

with RT = pv = p/γ = p/gρ.

For a subsonic flow M < 1, for a supersonic flow M > 1, and when w = a, M = 1.

In the presence of an adiabatic current, the temperature of a completely decelerated gas flow in any cross section at all in a combustion chamber and nozzle of the engine remains the same, i.e.,

$$T^* = T + \Delta T_{\text{дин}} = T + A \frac{w^2}{2 g c_p} = \text{const,} \quad *$$ (3.20)

where T and w are, respectively, the true temperature and velocity

————
*[$\Delta T_{\text{дин}}$ = ΔT_{din} = $\Delta T_{dinamicheskiy}$ = $\Delta T_{dynamic}$.]

of the gas flow before deceleration; and ΔT_{din} is the dynamic increase of the temperature as a consequence of the deceleration of the gas flow, attaining a considerable magnitude at velocities close to the speed of sound (for example, with a velocity $w = 10$ m/sec, $\Delta T_{din} = 0.05^\circ$C; when $w = 100$ m/sec, $\Delta T_{din} = 5.0^\circ$C, and when $w = 350$ m/sec, $\Delta T_{din} = 60^\circ$C).

Since from the equation of state for 1 kg of gas

$$c_p = \frac{k}{k-1} AR = \frac{k}{k-1} A \frac{p}{g\rho T}.$$

Expression (3.20) may be given the form

$$T^* = T + T \frac{k-1}{2} \frac{w^2}{k \frac{p}{\rho}} = T\left(1 + \frac{k-1}{2} M^2\right). \tag{3.20'}$$

We obtain an analogous expression for the temperature of the adiabatic deceleration of the gas flow in the inlet section of the engine nozzle:

$$T^* = T_k^* = T_k'\left(1 + \frac{k-1}{2} M_k^2\right), \tag{3.20''}$$

where T'_k is the true temperature of the gas flow at the entrance to the nozzle with a velocity w_k.

Since, in existing engines, gas velocity w_k at the entrance to the nozzle is about 60-200 m/sec, one may assume that the values of the true temperature T'_k and the temperature T_k determined by the thermodynamic calculation of the engine, and the temperature T^*_k of the adiabatic deceleration of the gas flow (corresponding to $w_k = 0$) differ very little in practice.

By comparing the right-hand parts of Eqs. (3.20'') and (3.20'), we learn the relationship of the true temperatures of the gas flow for any two sections of the engine nozzle being studied:

$$\frac{T}{T_k'} = \frac{1 + \frac{k-1}{2} M_k^2}{1 + \frac{k-1}{2} M^2}. \tag{3.21}$$

Since, for these nozzle sections in the presence of an adiabatic gas flow, the temperatures are associated with the pressures and densities by the well-known equations

$$\frac{T}{T_k} = \left(\frac{p}{p_k}\right)^{\frac{k-1}{k}} = \left(\frac{\rho}{\rho_k}\right)^{k-1}.$$

from which

$$\frac{p}{p_k} = \left(\frac{T}{T_k}\right)^{\frac{k}{k-1}} \quad \text{and} \quad \frac{\rho}{\rho_k} = \left(\frac{T}{T_k}\right)^{\frac{1}{k-1}}.$$

then Eq. (3.21) may be given the form

$$\frac{p}{p_k} = \left(\frac{1 + \frac{k-1}{2} M_k^2}{1 + \frac{k-1}{2} M^2}\right)^{\frac{k}{k-1}}$$

and

$$\frac{\rho}{\rho_k} = \left(\frac{1 + \frac{k-1}{2} M_k^2}{1 + \frac{k-1}{2} M^2}\right)^{\frac{1}{k-1}}. \tag{3.21'}$$

Analogously, one may obtain the relationships of the gas-flow parameters for the critical, outlet, or any of the cross sections of the engine nozzle.

If the velocity w_k of the gas flow at the outlet to the nozzle is practically negligible, and therefore one may accept $M_k = 0$, Eqs. (3.21) and (3.21') take the form

$$T/T'_k = T/T_k = (p/p_k)^{k-1/k} = 1/[1 + (k-1)/2M^2],$$

from which

$$M = \sqrt{\frac{2}{k-1}\left(\frac{T_k}{T} - 1\right)} = \sqrt{\frac{2}{k-1}\left[\left(\frac{p_k}{p}\right)^{\frac{k-1}{k}} - 1\right]};$$

and

$$\frac{p}{p_{\kappa}}=\frac{p}{p_{\kappa}}=\left(\frac{T}{T_{\kappa}}\right)^{\frac{k}{k-1}}=\left(\frac{1}{1+\frac{k-1}{2}M^2}\right)^{\frac{k}{k-1}}$$

$$\frac{\rho}{\rho_{\kappa}}=\frac{\rho}{\rho_{\kappa}}=\left(\frac{T}{T_{\kappa}}\right)^{\frac{1}{k-1}}=\left(\frac{1}{1+\frac{k-1}{2}M^2}\right)^{\frac{1}{k-1}}. \qquad (3.21'')$$

Applying these expressions to the critical section of the nozzle, i.e., using $M = M_{kr} = 1$ in them, we will obtain, accordingly,

$$\frac{T_{\kappa p}}{T_{\kappa}}=\frac{2}{k+1};\ \frac{p_{\kappa p}}{p_{\kappa}}=\left(\frac{2}{k+1}\right)^{\frac{k}{k-1}}\ \text{and}\ \frac{\rho_{\kappa p}}{\rho_{\kappa}}=\left(\frac{2}{k+1}\right)^{\frac{1}{k-1}}$$

For the combustion products of a ZhRD having a power of $k = 1.2$, we find $p_{kr} = 0.56 \ p_k$; $T_{kr} = 0.909T_k$, and $\rho_{kr} = 0.621\rho_k$.

The pressure of the adiabatically decelerated gas flow in any of the cross sections in the engine nozzle is determined by the adiabatic dependence of the gas's parameters, as well known from the techniques of thermodynamics:

$$p^*=p\left(\frac{T^*}{T}\right)^{\frac{k}{k-1}}\ \text{atm abs,}$$

where p and T are the static absolute pressure and the corresponding temperature in any nozzle cross section.

The introduction into ZhRD theory of the concepts of the parameters of adiabatic gas deceleration considerably simplifies calculations in a number of cases, since by this means the necessity of considering the change of the kinetic energy of the gas flow along the length of the combustion chamber and nozzle of the engine is excluded.

The relationship of the velocities of the gas flow for the two sections of the engine nozzle being studied are determined from the expression

$$\frac{M}{M_{\kappa}}=\frac{\frac{w}{\sqrt{kgRT}}}{\frac{w_{\kappa}}{\sqrt{kgR_{\kappa}T_{\kappa}}}}=\frac{w}{w_{\kappa}}\sqrt{\frac{T_{\kappa}}{T}},$$

i.e.,

$$\frac{w}{w_{\kappa}}=\frac{M}{M_{\kappa}}\sqrt{\frac{T}{T_{\kappa}}}. \qquad (3.22)$$

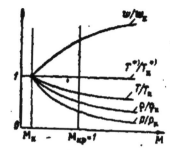

Fig. 3.8. Change of basic parameters of gas in the presence of an adiabatic current in an engine's geometrical chamber, dependent upon the Mach number.

since in the presence of an adiabatic gas flow $R_k = R'_k = $ const.

After replacing T/T'_k in this formula with the expression from Eq. (3.21), we finally obtain

$$\frac{w}{w_k} = \frac{M}{M_k} \sqrt{\frac{1 + \frac{k-1}{2} M_k^2}{1 + \frac{k-1}{2} M^2}} \cdot \text{(3.23)} \qquad (3.23)$$

Figure 3.8 shows the dependence of the parameters of the gas flow along the length of the geometrical nozzle on the Mach number.

In some cases design formulas are simplified if the parameters of the gas condition are determined in the form of dependences not on the Mach number, but on velocity factor $\lambda = w/w_{kr} = w/a_{kr}$.

The dependences of the change of temperature, pressure, and other parameters of the gas on the velocity factor λ are expressed by the following well-known formulas:

$$\frac{T}{T_k} = 1 - \frac{k-1}{k+1} \lambda^2;$$

$$\frac{p}{p_k} = \left(1 - \frac{k-1}{k+1} \lambda^2\right)^{\frac{k}{k-1}};$$

$$\frac{\rho}{\rho_k} = \left(1 - \frac{k-1}{k+1} \lambda^2\right)^{\frac{1}{k-1}};$$

$$\sqrt{\frac{k+1}{k-1}\left[1 - \left(\frac{p}{p_k}\right)^{\frac{k-1}{k}}\right]} = \sqrt{\frac{k+1}{k-1+\frac{2}{M^2}}};$$

$$M = \lambda \sqrt{\frac{2}{k+1-(k-1)\lambda^2}}.$$

Since in the calculations of a <u>ZhRD</u> a one-dimensional gas flow is being investigated, then, regardless of the nozzle profile, the only variable magnitude applicable to the nozzle is the area F of its flow-through section.

Because pressure p is the basic parameter characterizing the change in the state of the gas stream along the length of the nozzle, we ordinarily determine the dependence of the change of p on F, and then compute the other parameters of the gas corresponding to this pressure.

The relationship of the arbitrary and critical cross sections of the engine nozzle, depending on the pressure of the gas flow, is determined from the equation of continuity of flow, written in conformity to these nozzle sections:

$$G = F w \rho = F_{\kappa p} w_{\kappa p} \rho_{\kappa p} = \text{const,} \qquad (3.24)$$

i.e.,

$$\frac{F}{F_{\kappa p}} = \frac{w_{\kappa p}}{w} \cdot \frac{\rho_{\kappa p}}{\rho}.$$

After substitution in this equation of the well-known expressions for w_{kr}, w, ρ_{kr}, and ρ and the corresponding changes, we finally obtain

$$f = \frac{F}{F_{\kappa p}} = \left[\frac{2}{k+1}\right]^{\frac{1}{k-1}} \left(\frac{p_\kappa}{p}\right)^{\frac{1}{k}} \sqrt{\frac{k-1}{k+1}\left[1 - \left(\frac{p_\kappa}{p}\right)^{\frac{k-1}{k}}\right]} =$$

$$= \frac{1}{M}\left(\frac{1 + \frac{k-1}{2} M^2}{\frac{k+1}{2}}\right)^{\frac{k+1}{2(k-1)}} = \frac{1}{\lambda}\left(\frac{\frac{2}{k+1}}{1 - \frac{k-1}{k+1}\lambda^2}\right)^{\frac{1}{k-1}}. \qquad (3.25)$$

This equation gives the necessary connection between the magnitude F of the nozzle's (flow-through) cross section and the pressure of the gas flow in the given section (with given values of F_{kr} and p_k) and also may be used to determine the outlet section F_v of the nozzle at a gas pressure p_v.

The connection between F and p is not direct, but, having established the necessary relationship p_k/p, we may determine the unknown relationship of areas $f = F/F_{kr}$ (Fig. 3.9) in accordance with the equation given.

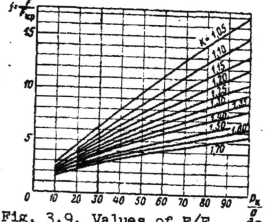

Fig. 3.9. Values of F/F_{kr}, dependent on relationship p_k/p_v and adiabatic exponent $\underline{k}$ or polytropic exponent $\underline{n}$.

Assuming a series of values for gas-stream pressures along the length of the engine chamber nozzle, one may compute the other gas parameters corresponding to them:

$$T = T_\kappa\left(\frac{p}{p_\kappa}\right)^{\frac{k-1}{k}}; \quad \rho = \rho_\kappa\left(\frac{p}{p_\kappa}\right)^{\frac{1}{k}};$$

$$\gamma = \gamma_\kappa\left(\frac{p}{p_\kappa}\right)^{\frac{1}{k}}; \quad v = v_\kappa\left(\frac{p_\kappa}{p}\right)^{\frac{1}{k}};$$

where

$$w = \lambda w_{\kappa p} = \sqrt{\frac{k+1}{k-1}\left[1-\left(\frac{p}{p_\kappa}\right)^{\frac{k-1}{k}}\right]} =$$

$$= \sqrt{2g\frac{k}{k-1}R_\kappa T_\kappa\left[1-\left(\frac{p}{p_\kappa}\right)^{\frac{k-1}{k}}\right]},$$

$$w_{\kappa p} = \sqrt{2g\frac{k}{k+1}R_\kappa T_\kappa} = \sqrt{gkR_\kappa T_{\kappa p}}.$$

Having the basic geometrical dimensions of the nozzle and having drawn it in the corresponding scale, according to the expression $d = \sqrt{4F/\pi}$ one may determine the sections which will correspond to the computed values of the parameters of the gas flow and plot the curves of their change along the length of the chamber nozzle (Fig. 3.10).

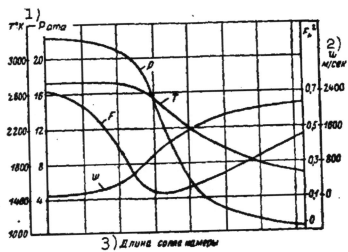

Fig. 3.10. Change of gas-flow parameters along the length of an engine chamber nozzle. 1) p, atm abs; 2) w, m/sec; 3) length of chamber nozzle.

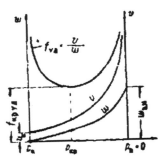

Fig. 3.11. Dimensionless area of an engine's geometrical nozzle, dependent on Mach number in the presence of an adiabatic gas current.

The equations derived here for an adiabatic gas current are also applied for isentropic and polytropic processes of gas expansion in the engine nozzle upon condition that the adiabatic exponent k in these equations is replaced by the isentropic or polytropic exponent n.

Ordinarily the polytropic exponent $n \approx 1.10-1.20$ and $F_v/F_{kr} \approx 3-4$.

Upon increase of k, the relationship F_v/F_{kr} is decreased, which shows the advantage of using fuels whose combustion products have a greater value of k.

The greater the pressure difference p_k/p_v of the gases in the nozzle, the greater the length of the nozzle at a given flare angle, the nozzle surface, and its weight.

In the presence of a subsonic current, gas velocity w is increased more intensively than specific volume v, and just the opposite in the presence of a supersonic current. Therefore the area of the

96

nozzle channel in the subsonic region must be decreased, and increased in the supersonic region, which is obvious from the equation of continuity of the gas stream $F = Gv/w$ (Fig. 3.11).

Consequently, the critical section of the engine nozzle is the boundary where the character of the geometrical influence on the gas flow changes to the opposite sign.

In an actual process, heat $dQ > 0$, and the gas friction work $dL > 0$; therefore the equality of velocities $w = a$ is established where $dF > 0$, i.e., in the expanding portion of the nozzle. This means that the gases' velocity in the narrowest section of the nozzle is less than the speed of sound.

Example 2. Construct the curves of the adiabatic change of gas flow parameters v, T, w, F, and M, depending on the length of the nozzle, if $p_k = 21$ atm abs; $p_v = 1.03$ atm abs; $T_k = 2200°K$; $G_g = G_s = 1$ kg/sec; $v_k = 0.373$ m^3/kg; $R_k = 35.3$ kg-m/kg-°C, and $k = 1.3$.

Solution.

We assume throughout the nozzle length, a series of values for gas pressure: p = 17.5, 14.0, 11.5, 7.0 and 4.2 atmospheres absolute and for each of them we determine the area of the cross section of the nozzle.

With a pressure p = 17.5 atm abs we have:

1) specific volume of gases

$$v = v_k \left(\frac{p_k}{p}\right)^{\frac{1}{k}} = 0.373 \left(\frac{21}{17.5}\right)^{1/1.3} = 0.43 \text{ m}^3/\text{kg};$$

2) gas temperature

$$T = T_k \left(\frac{p}{p_k}\right)^{\frac{k-1}{k}} = 2220 \left(\frac{17.5}{21}\right)^{\frac{1.3-1}{1.3}} = 2100° \text{ K};$$

3) exit velocity

$$w = \sqrt{2g \frac{k}{k-1} R_k T_k \left[1 - \left(\frac{p}{p_k}\right)^{\frac{k-1}{k}}\right]} =$$

$$= \sqrt{2 \cdot 9,81 \cdot 35,3 \cdot 2220 \left[1 - \left(\frac{17,5}{21} \right)^{\frac{1,3-1}{1,3}} \right]} = 530 \text{ m/sec};$$

4) area of the nozzle cross section

$$F = \frac{G_s v}{w} = \frac{10 \cdot 0,43}{530} = 0,000812 \text{ } м^2 = 8,12 \text{ } см^2;$$

5) relationship of local velocity to the speed of sound

$$M = \frac{w}{\sqrt{kgRT}} = \frac{530}{\sqrt{9,81 \cdot 1,3 \cdot 35,3 \cdot 2100}} = 0,53.$$

Analogously, we determine the parameters of the gas flow at the other chosen values of pressure and reduce the results of the calculations in Table 3.2.

TABLE 3.2

1) p кг/см²	2) v м³/кг	T °K	3) w м/сек	w/v	F см²	M
$p_к$=21	0,37	2200	0	0	—	0
17,5	0,43	2100	530	1230	3,12	0,53
14,0	0,51	2020	775	1520	6,50	0,81
$p_{кр}$=11,5	0,60	1930	980	1550	6,32	1,00
7,0	0,88	1720	1220	1390	7,10	1,42
4,2	1,30	1530	1440	1110	8,82	1,79
$p_в$=1,03	3,80	1105	1830	482	20,20	2,59

1) p, kg/cm²; 2) v, m³/kg; 3) w, m/sec.

Example 3. Determine the basic geometrical dimensions of an engine chamber nozzle if G_s = 13.54 kg/sec; p_k = 20 atm abs; T_k = 2787°K; γ_k = 2.023 kg/m³; w_v = 2174 m/sec; γ_v = 0.153 kg/m³; R = 35.25 kg-m/kg-°C, and $\underline{n}$ = 1.16.

Solution.

1. Parameters of the gas mixture in the critical section of the nozzle

$$T_{кр} = \frac{2T_к}{n+1} = \frac{2 \cdot 2787}{1,16+1} = 2580° \text{ K};$$

$$p_{кр} = p_к \left(\frac{2}{n+1} \right)^{\frac{n}{n-1}} = 20 \left(\frac{2}{1,16+1} \right)^{\frac{1,16}{1,16-1}} = 11,43\,kg/cm^3;$$

$$\gamma_{кр} = \frac{p_{кр}}{RT_{кр}} = \frac{11,43 \cdot 10^4}{35,27 \cdot 2580} = 1,257\,kg/cm^3;$$

$$w_{кр} = \sqrt{gnRT_{кр}} = \sqrt{9,81 \cdot 1,16 \cdot 35,25 \cdot 2580} = 1017\,m/sec.$$

2. Basic geometrical dimensions of the nozzle:

$$F_{кр} = \frac{G_s}{w_{кр}\gamma_{кр}} = \frac{13,54 \cdot 10^4}{1017 \cdot 1,257} = 105,9\,cm^2;$$

$$d_{кр} = 1,128\sqrt{F_{кр}} = 1,128\sqrt{105,9} = 13,17\,cm = 131,7\,mm;$$

$$F_s = \frac{G_s}{w_s\gamma_s} = \frac{13,54 \cdot 10^4}{2174 \cdot 0,153} = 407,1\,cm^2;$$

$$d_s = 1,128\sqrt{F_s} = 1,128\sqrt{407,1} = 25,6\,cm = 256\,mm.$$

Example 4. Determine the theoretical specific thrust at ground level and the specific areas of the critical and outlet sections of the nozzle of an engine chamber, operating on kerosene and nitric acid of 96% weight concentration, when $\alpha = 0.78$; $p_k = 40$ atm abs, and $p_v = 0.8$ atm abs, if the thermodynamic calculations produce the following theoretical parameters of the gases in the combustion chamber: $T_k = 2900°K$, $R_k = 34$ kg-m/kg-°C, and k = 1.16.

Solution.

1. Density of the gas in the combustion chamber

$$\gamma_к = \frac{p_к}{R_к T_к} = \frac{40 \cdot 10^4}{34 \cdot 2900} = 4,06\,kg/m^3.$$

2. Parameters of the gas in the outlet section of the chamber nozzle:

a) temperature

$$T_s = T_к \left(\frac{p_s}{p_к} \right)^{\frac{k-1}{k}} = 2900 \left(\frac{0,8}{40} \right)^{\frac{1,16-1}{1,16}} = 2900 \cdot 0,5839 = 1690°\,K;$$

b) density

$$\gamma_s = \gamma_к \left(\frac{T_s}{T_к} \right)^{\frac{1}{k-1}} = 4,06 \left(\frac{1690}{2900} \right)^{\frac{1}{1,16-1}} = 4,06 \cdot 0,052 = 0,21\,kg/m^3;$$

c) velocity factor

$$\lambda_s = \sqrt{ \frac{k+1}{k-1} \left[1 - \left(\frac{p_s}{p_к} \right)^{\frac{k-1}{k}} \right] } = \sqrt{ \frac{1,16+1}{1,16-1} \left[1 - \left(\frac{0,8}{40} \right)^{\frac{1,16-1}{1,16}} \right] } = 2,37;$$

d) exit velocity

$$w_s = \lambda_s a_{\kappa p} = \lambda_s \sqrt{2g \dfrac{k}{k+1} R_\kappa T_\kappa} =$$

$$= 2{,}37 \sqrt{2 \cdot 9{,}81 \dfrac{1{,}16}{1{,}16+1} 34 \cdot 2900} = 2418 \text{ m/sec};$$

e) relative area of the outlet section of the chamber nozzle

$$f_s = \dfrac{F_s}{F_{\kappa p}} = \dfrac{1}{\lambda_s}\left(\dfrac{\dfrac{2}{k+1}}{1 - \dfrac{k-1}{2}\lambda_s^2}\right)^{\frac{1}{k-1}} =$$

$$= \dfrac{1}{2{,}37}\left(\dfrac{\dfrac{2}{1{,}16+1}}{1 - \dfrac{1{,}16-1}{2} \cdot 2{,}37^2}\right)^{\frac{1}{1{,}16-1}} = 5{,}024;$$

f) specific area of the outlet section of the chamber nozzle

$$f_{y\lambda.s} = \dfrac{1}{w_s \gamma_s} = \dfrac{1}{2418 \cdot 0{,}21} = 0{,}001974 \text{ m}^2/\text{kg} = 19{,}74 \text{ cm}^2/\text{kg};$$

g) specific area of the critical section of the chamber nozzle

$$f_{y\lambda.\kappa p} = f_s f_{y\lambda.\kappa p} = 5{,}024 \cdot 19{,}74 = 3{,}93 \text{ cm}^2/\text{kg}.$$

3. Specific thrust of the engine chamber

$$p_{y\lambda.s} = \dfrac{P_{\scriptscriptstyle H}}{G_s} = \dfrac{\dfrac{G_s}{g}w_s + F_s(p_s - p_s)}{G_s} = \dfrac{w_s}{g} + \dfrac{F_s}{G_s}(p_s - p_s) =$$

$$= \dfrac{w_s}{g} + f_{y\lambda.s}(p_s - p_s) = \dfrac{2418}{9{,}81} + 19{,}34\,(0{,}8 - 1) = 243{,}052 \text{ kg-sec/kg}.$$

Chapter IV

ZhRD OPERATING REGIMES

In use, a ZhRD operates under various internal regimes and external conditions.

Control of the magnitude of engine thrust is usually accomplished by a change in the per-second rate of fuel flow into the combustion chamber and by a pressure change in the latter. By this means, not only is the thrust changed, but the engine operating economy as well.

With a change in the conditions of the surrounding medium, the basic parameters of the ZhRD are also changed, since, in this case, the back pressure behind the nozzle is reduced, which, other things being equal, is also reflected in the magnitudes of the engine's absolute and specific thrusts. In designing ZhRD weapons, one must know the basic parameters of the engine (absolute and specific thrusts) under rated and various nonrated static operating regimes at ground or sea level or in flight conditions. The basic factors determining the operating regime of the engine and upon which absolute and specific thrust depend, i.e., pressure in the combustion chamber, flight altitude, etc., are called engine characteristics.

In this chapter there is a discussion of ZhRD characteristics under possible internal regimes and conditions of operation, and methods are laid down for thus determining absolute and specific thrust as well as specific fuel consumption and other parameters for engine design.

Furthermore, a brief analysis of the factors affecting the magnitude of specific thrust is given, and suggestions are made for increasing engine operating economy.

This material facilitates determination of the optimum parameters and selection of an expedient operating regime for the engine being designed.

SECTION 1. RATED AND NONRATED ENGINE OPERATING REGIMES

The operational condition of a ZhRD is characterized by a set of parameters determining its efficiency, absolute and specific thrusts, and the heat, static, and dynamic loads on individual engine elements; the accepted designation of this state is the operating regime of the engine and the parameters of the operating process in the engine which correspond to it, such as fuel flow rate per second and composition factor, pressure in the combustion chamber and in the nozzle outlet section, and others, as engine operating-regime parameters.

By the use of a ZhRD thermodynamic rating, one may evaluate the basic engine characteristics only under certain rated operational conditions alone.

In operational conditions, absolute and specific engine thrusts may change subject to the internal operating regime of the engine (from a change of gas pressure in the combustion chamber as a result of a change of the per-second fuel flow rate in it or the weight relationship of its components) and external conditions (flight altitude and velocity or pressure of the surrounding medium).

If a ZhRD operates at a rated per-second rate of fuel flow into the combustion chamber (i.e., rated pressure in the chamber), the regime is referred to as rated, and if the per-second fuel flow rate to the combustion chamber is greater or smaller than rated, the regime is referred to as <u>nonrated</u>.

The ratio of gas pressure p_k in the combustion chamber at a given engine operating regime to the gas pressure p_v in the outlet section of the nozzle of this chamber is called the <u>gas expansion ratio in the nozzle</u>.

If, in the chamber nozzle, the gases expand up to the ambient air pressure $(p_v = p_a)$, the gas expansion ratio in the nozzle and the nozzle operating regime are referred to as <u>optimum</u>; if the gases in the nozzle exceed or fail to attain the ambient air pressure $(p_v \gtrless p_a)$, the gas expansion ratio in the nozzle and the nozzle operating regime is called <u>nonoptimum</u>.

A change in the operating regime of an engine nozzle may be caused by a change in the per-second fuel flow rate to the combustion chamber relative to the rated value, or by a change in flight altitude (atmospheric pressure). An engine nozzle may operate on an optimum regime only if the values of altitude and flight speed of the missile and the per-second rate of fuel flow to the combustion chamber are constant. Under operating conditions the chamber nozzles of missile engines usually operate at nonoptimum regimes of overexpansion and underexpansion of the gases.

The most advantageous engine operating regime is the regime rated at optimum gas expansion in the nozzle. This operating regime is called <u>optimum</u>, since it corresponds to the greatest values of actual efficiency and specific thrust relative to their values when the chamber nozzles operate under nonoptimum regimes (underexpansion and overexpansion of the gases in the nozzle). Numerical values of operating parameters are usually given for the engine's work at this regime.

Existing <u>ZhRD</u> may be divided into two groups depending on the operating regime of the engine nozzle at ground or sea level:

1) _low-altitude_ engines, in which, during normal operation at ground or sea level, gas pressure in the outlet section of the nozzle is equal to or greater than atmospheric pressure, i.e., when $p_v = p_a$;

2) _high-altitude_ engines, in which, during normal operation at ground or sea level, gas pressure in the outlet section of the nozzle is less than the pressure of the atmospheric air, i.e., when $p_v < p_a$.

A high-altitude engine may operate on a regime of optimal gas expansion in the nozzle only at a rated altitude H_{rasch},* where an equality between p_v and p_a is attained.

The basic characteristic of a high-altitude engine is the magnitude of the gas pressure p_v in the nozzle outlet section. Consequently, by a rated altitude of an engine nozzle, we understand that rated flight altitude where $p_v = p_a$.

Depending on the rated altitude of the nozzle, the accepted subdivision of liquid-fuel rocket engines is into engines of _moderate altitude_ (when $p_v > 0.6$ atm abs) and of _high altitude_ (when $p_v < 0.6$ atm abs). The engines of the last stages of multistage missiles are of the high-altitude type.

SECTION 2. OPERATIONAL ENGINE OPERATING REGIMES

With regard to the per-second fuel flow rate into the combustion chamber, we may distinguish the following characteristic operating regimes (thrusts) of an engine.

1. _Starting_, constituting the operation of an engine in the period from the instant fuel is ignited in the combustion chamber until the engine reaches its operating regime.

2. _Nominal thrust regime_, corresponding to continuous normal load of the engine at a nominal (specific) rate of propellant consumption

*[$H_{pacu} = H_{rasch} = H_{raschetnaya} = H_{rated}$.]

by the engine chamber.

3. **Maximum thrust regime**, corresponding to a brief maximum engine load at the maximum permissible flow rate intensity of its combustion chamber and an increase in the specific propellant (fuel) consumption in comparison to the nominal regime.

Other conditions being equal, such forced engine operation is combined with a deterioration in engine operating economy because of the inadequate volume of the combustion chamber or the nonrated per-second fuel flow rate into the combustion chamber, and other reasons.

A temporary increase in engine thrust relative to the rated nominal thrust may necessitate an increase in the weapon's rate of climb, decrease of the length of the aerodynamic vehicle's run at blastoff, or a temporary increase of maximum flight speed.

In operation, a ZhRD seldom operates at a maximum thrust regime. Usually, an engine operates with a thrust less than maximum the greater part of the time.

4. **Minimum thrust regime**, corresponding to the minimum permissible load of the engine, below which its operation becomes unstable.

The value of minimum thrust depends on the engine design and designation and sometimes amounts to about 10-30% of nominal thrust.

5. **Operational regime**, corresponding to engine loads in the range from minimum to established maximum thrust magnitude, within which stable engine operation is ensured.

The duration of ZhRD operation at pertinent regimes is established by the engine designer.

The above classification of operating (thrust) regimes of an engine may have particular significance for main engines and aircraft vernier engines. Boosters have only one operating regime.

6. **Stable operating regime** of the engine, corresponding to the

load at which the fuel combustion process in the chamber occurs without anomalous pulsations in gas pressure.

SECTION 3. FACTORS AFFECTING THE INTENSITY OF ZhRD THRUST

The result of all hydrodynamic pressure forces operating on an engine chamber in the presence of outflow of gases from it to the surrounding medium is called the thrust of a ZhRD.

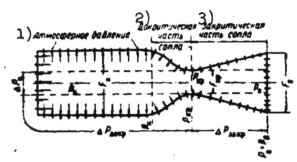

Fig. 4.1. Definition of the thrust of a ZhRD as the force resulting from the influence of gas pressure on the internal surface and air pressure on the external surface of the engine chamber. 1) Atmospheric pressure; 2) subcritical part of the nozzle; 3) hypercritical part of the nozzle.

The thrust of an engine, neglecting atmospheric air pressure (when the engine is operating in space), is dependent upon the following three forces of gas pressure working on the combustion chamber and nozzle in the direction of the nozzle-combustion chamber axis (Fig. 4.1):

1) the unbalanced force of gas pressure on the surface of the combustion chamber head, equal to the pressure p_k produced in the chamber on the area of the head, in a magnitude equal to the area F_{kr}, of the critical section of the nozzle, i.e.,

$$\Delta P_k = p_k F_{kr};$$ (4.1)

2) a force equal to the difference of the gas pressure forces act-

ing on the remaining part of the combustion chamber head (area $F_k -$
$- F_{kr}$), and the subcritical part of the nozzle which has the same area
in the direction of the engine's axis:

$$\Delta P_{dokr} = (p_k - p_{rez.dokr})(F_k - F_{kr});^* \qquad (4.2)$$

3) the unbalanced force of the gas pressure on the hypercritical
part of the nozzle which has the area $F_v - F_{kr}$ in the direction of the
engine's axis

$$\Delta P_{zakr} = p_{rez.zakr}(F_v - F_{kr}),^{**} \qquad (4.3)$$

that is,

$$P_p = \Delta P_k + \Delta P_{dokr} + \Delta P_{zakr}, \qquad (4.4)$$

where p_k is the absolute gas pressure in the combustion chamber (pressure on the chamber head) in atm abs; F_k and F_v are the area of the
greatest cross sections in the combustion chamber and the outlet section of the nozzle, respectively; $p_{rez.dokr}$ and $p_{rez.zakr}$ are the resultant absolute gas pressures on the subcritical and hypercritical
nozzle sections, respectively, in atm abs.

Gas-pressure forces on the engine chamber in a direction perpen-
dicular to its axis balance each other in a manner analogous to the
forces of air pressure on an engine during its operation in the atmosphere (see Fig. 4.1).

The resultant of all axial forces caused by the pressure of the
liquid in the cooling duct of the engine chamber is usually negligible.

The absolute thrust of an engine chamber when operating in the atmosphere is expressed by the equation

---

*[$p_{рез.докр} = p_{rez.dokr} = p_{rezul'tiruyushchaya.dokriticheskaya} =$
$= p_{resultant.subcritical}.$]
**[$p_{рез.закр} = p_{rez.zakr} = p_{rezul'tiruyushchaya.zakriticheskaya} =$
$= p_{resultant.hypercritical}.$]

- - -

$$P_H = P_p - F_v p_a, \tag{4.5}$$

where $F_v p_a$ is the force of the counter pressure of the atmospheric air on the gas flow in the outlet section of the engine chamber nozzle.

Thrust magnitude P_p depends only on the internal conditions of the operation of a given engine, characterized by the pressure p_k in the combustion chamber, and the term $F_v p_a$ depends only on the external conditions of the engine's operation, i.e., on the pressure p_a of the surrounding medium. When the engine is operating in space, where $p_a = 0$, the term $F_v p_a = 0$.

The thrust of an existing engine may be determined on a test stand under fixed atmospheric conditions.

SECTION 4. THE ABSOLUTE THRUST OF AN ENGINE

The absolute thrust of a liquid-fueled rocket engine chamber usually consists of:

1) <u>the dynamic thrust component</u>, equal to the change of momentum of the combustion products of the per-second fuel flow rate to the combustion chamber of the engine:

$$P_{дин} = \frac{G_t}{g} \cdot w \text{ kg}; \tag{4.6}$$

2) <u>the static thrust component</u>, equal to the force of the pressure on the engine of the fuel combustion products which are expanding behind the nozzle:

$$P_{ст} = F_a (p_a - p_a) \text{ kg,}* \tag{4.7}$$

that is,

$$P_z = P_{дин} + P_{ст} = \frac{G_t}{g} w_a + F_a (p_a - p_a) =$$
$$= \frac{G_t}{g} w_a + F_a p_a \left(1 - \frac{p_a}{p_a}\right) = \frac{G_t}{g} w_{эфф}. \tag{4.8}$$

_____

*[P_{CT} = P_{st} = $P_{staticheskiy}$ = P_{static}.]

This expression is called the general thrust equation of an engine. This equation may be given other expressions, often encountered in the literature, if you consider that

i.e.,

$$w_s = 91{,}53 \sqrt{H_s \bar{\eta}_t}; \quad F_s = G_s/w_s\gamma_s = G_s v_s/w_s \text{ и } p_s v_s = R_s T_s,$$

$$P_s = \frac{G_t}{g} 91{,}53 \sqrt{H_s \bar{\eta}_t} + F_s (p_s - p_s) =$$

$$= 9{,}33 G_t \sqrt{H_s \bar{\eta}_t} + F_s (p_s - p_s) = \frac{G_t}{g} w_{эф.u}; \quad (4.8')$$

$$P_s = \frac{G_t}{g} w_s + \frac{G_t}{w_s \gamma_s} (p_s - p_s) = \frac{G_t}{g} \left(w_s + g \frac{p_s - p_s}{w_s \gamma_s} \right);$$

$$P_s = \frac{G_t}{g} w_s + \frac{G_t v_s}{w_s} p_s \left(1 - \frac{p_s}{p_s} \right) = \frac{G_t}{g} \left[w_s + \frac{g R_s T_s}{w_s} \left(1 - \frac{p_s}{p_s} \right) \right].$$

The latter two equations permit the determination of an engine's thrust independently of its geometrical dimensions.

The thrust equation may also be given this form:

$$P_u = \frac{G_t}{g} w_s + F_s p_s - F_s p_s = F_s p_s \left(1 + \frac{G_t w_s}{g F_s p_s} \right) - F_s p_s =$$

$$= F_s p_s \left(1 + \frac{F_s w_s \gamma_s w_s}{g F_s p_s} \right) - F_s p_s = F_s p_s \left(1 + \frac{n}{n} \frac{w_s^2}{g} \frac{\gamma_s}{p_s} \right) - F_s p_s =$$

$$= F_s p_s \left(1 + n \frac{w_s^2}{n g R_s T_s} \right) - F_s p_s$$

or, finally,

$$P_s = F_s p_s (1 + n M_s^2) - F_s p_s = \frac{G_t}{g} w_{эф.u}, \quad (4.8'')$$

where $\underline{n}$ is the exponent of polytropic gas expansion in the nozzle, $M_v = w_v/a_v$ is the ratio of the gas exhaust velocity w_v in the outlet section of the nozzle to the local speed of sound a_v (Mach number).

The general thrust equation may be used for calculation of engine characteristics.

In the chamber of an engine operating on a given fuel when $p_v/p_a \geq 0.3$-0.4, the value of p_k/p_v is not dependent on G_s or p_k; the pressure p_v does not depend on p_a, but upon p_k; velocity w_v does not depend on p_k and p_a, but upon p_k/p_v (on the dimensions of the chamber nozzle), i.e., for a given chamber at standard operating conditions, the magnitude of w_v is constant.

For calculating flow rate characteristics, the general thrust equation must be reduced to another form.

If, in the general thrust equation of an engine

$$P_s = \frac{G_s}{g} w_s + F_s(p_s - p_a)$$

we substitute

$$G_s = \frac{F_{\kappa p} w_{\kappa p}}{v_{\kappa p}}, \text{ then } w_s = \sqrt{2g \frac{n}{n-1} p_\kappa v_\kappa \left[1 - \left(\frac{p_s}{p_\kappa}\right)^{\frac{n-1}{n}}\right]},$$

$$w_{\kappa p} = \sqrt{2g \frac{n}{n+1} p_\kappa v_\kappa} \text{ and } v_{\kappa p} = v_\kappa \left(\frac{n+1}{2}\right)^{\frac{1}{n-1}},$$

we will obtain

$$P_s = \frac{F_{\kappa p} w_{\kappa p}}{v_{\kappa p}} \frac{w_s}{g} + F_s (p_s - p_a) =$$

$$= p_\kappa F_{\kappa p} \sqrt{\frac{2n^2}{n-1} \left(\frac{2}{n+1}\right)^{\frac{n+1}{n-1}} \left[1 - \frac{p_s}{p_\kappa}^{\frac{n-1}{n}}\right]} + F_s (p_s - p_a). \qquad (4.8''')$$

This equation shows that the absolute thrust of an engine is proportional to the pressure p_k in the combustion chamber and the area F_{kr} of the critical section of the nozzle.

When an engine is operating in space, where $p_a = 0$ and therefore $F_v p_a = 0$, Eqs. (4.8) through (4.8''') are expressed as follows:

$$\left. \begin{aligned} &P_n = 9.33 G_s \sqrt{H_s T_\kappa} + F_s p_s = \frac{G_s}{g} w_{s\phi.n}; \\ &P_n = \frac{G_s}{g} w_s + F_s p_s; \\ &P_n = \frac{G_s}{g} \left(w_s + \frac{g R_s T_s}{w_s}\right); \\ &P_s = F_s p_s (1 + n M_s^2); \\ &P_n = p_\kappa F_{\kappa p} \sqrt{\frac{2n^2}{n-1} \left(\frac{2}{n+1}\right)^{\frac{n+1}{n-1}} \left[1 - \left(\frac{p_s}{p_\kappa}\right)^{\frac{n-1}{n}}\right]} + F_s p_s \end{aligned} \right\} \qquad (4.9)$$

When changes occur in combustion chamber pressure p_k, altitude H, and flight speed V, the absolute thrust of a given engine changes, since

$$p_v = f(p_k) \text{ and } p_a = f'(H,V).$$

With a constant G_s and the same nozzle dimensions, P_{din} is not changed, and P_{st} may be changed only by a change of p_a. When $p_v = p_a$ the value of $P_{st} = 0$, when $p_v > p_a$ it is positive, and when $p_v < p_a$, it is negative (Fig. 4.2).

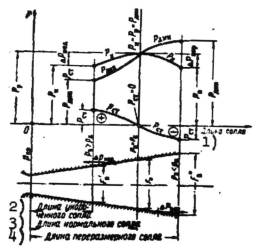

Fig. 4.2. Absolute thrust of an engine as a function of nozzle length at rate of fuel flow to combustion chamber. 1) Nozzle length; 2) length of a shortened nozzle; 3) normal nozzle length; 4) extended nozzle length.

Expansion of the gases in an engine chamber nozzle up to the pressure of the surrounding medium is possible only when the nozzle is of normal length.

The absolute thrust of an engine when the nozzle is operating on an optimum (rated) regime is expressed as follows

$$P_{опт} = P_p = P_{дин} = \frac{G_t}{g} w_v. \quad *$$

(4.10)

*[$P_{опт} = P_{opt} = P_{optimal'nyy} = P_{optimum}$;
$P_p = P_r = P_{raschetnyy} = P_{rated}$.]

The optimum operating regime of a ZhRD nozzle is the most advantageous since, other conditions being equal, the engine develops maximum thrust under this regime.

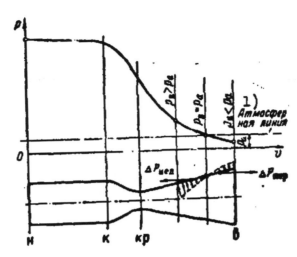

Fig. 4.3. Change of gas pressure on the surface of a shortened, normal, and lengthened nozzle. 1) Atmosphere line.

An engine's thrust P_n when the nozzle operates at a regime of underexpansion or overexpansion of gases is always less than the engine's thrust P_{opt} obtained at rated optimum operating regime of the combustion chamber and nozzle, according to the magnitudes (Fig. 4.2 and 4.3):

$$\Delta P_{нед} = (F_a - F_a') p_{рез.нед} \quad *$$

and

$$\Delta P_{пер} = (F_a' - F_a) p_{рез.пер} \quad ** \qquad\qquad (4.11)$$

where $p_{rez.ned}$ and $p_{rez.per}$ are the resultant true gas pressures on the surface of the shortened and lengthened parts, respectively, of the engine chamber nozzle.

---

*[$\Delta P_{нед}$ = ΔP_{ned} = $\Delta P_{nedorasshirenie}$ = $\Delta P_{underexpansion}$;
 $p_{рез.нед}$ = $p_{resultant.underexpansion}$.]
**[$\Delta P_{пер}$ = ΔP_{per} = $\Delta P_{pererasshirenie}$ = $\Delta P_{overexpansion}$;
 $p_{рез.пер}$ = $p_{resultant.overexpansion}$.]

Therefore, the absolute thrust of an engine chamber, when the nozzle is operating on a nonoptimum regime, will be:

1) in case of underexpansion of the gases in the nozzle

$$P_a = P_{ont} - \Delta P_{ned},$$

from whence $P_{opt} = P_n + \Delta P_{ned}$;

2) in case of overexpansion of the gases in the nozzle

$$P_a = P_{ont} - \Delta P_{per},$$

from whence $P_{opt} = P_n + \Delta P_{per}$.

Consequently, at given values of G_s and p_k, shortening or lengthening of a given engine nozzle will lead to a change in the magnitudes of p_v and w_v, and P_{din} and P_{st}. For example, when a nozzle of normal length is shortened, p_v increases and P_{din} decreases as the consequence of some underexpansion of the gases in the nozzle, and P_{st} increases because of the increase in the pressure difference of the gases behind the nozzle, as a result of which the engine's thrust is, accordingly, decreased relative to the rated optimum magnitude P_{opt}.

Change of an engine's absolute thrust, depending on the change of nozzle length, corresponds to a definite change of altitude H or flying speed V.

The thrust of a low-altitude engine, when operating at a certain altitude H, is expressed by the equation

$$P_N = P_{ont} + F_a (\rho_a - \rho_a). \qquad (4.12)$$

A low-altitude engine in space develops approximately 10 to 20% greater thrust than at ground level.

Since the engine of a long-range missile usually operates in the rarefied layers of the atmosphere during the greater part of the time, it is expedient to make it a high-altitude engine, i.e., with a relatively lengthened nozzle at ground or sea level (when $p_v < p_a$).

Such a high-altitude engine, when operating at ground level or at

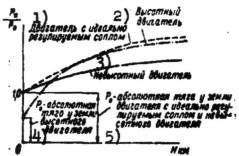

Fig. 4.4. Change of absolute thrust of low-altitude and high-altitude ZhRD, depending on flight altitude H. 1) Engine with ideal regulating nozzle; 2) high-altitude engine; 3) low-altitude engine; 4) P_0 — absolute thrust of a high-altitude engine at ground level; 5) P_0 — absolute thrust at ground level of an engine with ideal regulating nozzle and low-altitude engine.

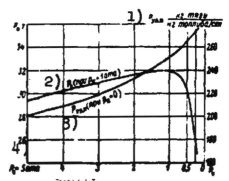

Fig. 4.5. Absolute thrust of an engine chamber operating at ground level and its specific thrust in space, depending on p_v when p_k is constant. 1) $P_{ud.p}$, kg thrust/kg fuel per sec; 2) P_0 (when p_a = 1 atm abs); 3) $P_{ud.p}$ (when p_a = 0); 4) P_v = 5 atm abs.

low altitudes, will develop less thrust than a low-altitude engine (with a normal nozzle length). However, to make up for that, as it rises in altitude the thrust of such an engine becomes, other things being equal, greater than the thrust of a low-altitude engine within a short time (Fig. 4.4).

A missile with such a high-altitude engine, other things being equal, will have a relatively great range. For this reason, at the present time high-altitude engines are made for long-range missiles and for some antiaircraft missiles. The design altitude for an engine being designed is usually chosen with due regard to the missile's type and tactical designation. The absolute thrust of a high-altitude engine is determined according to the general thrust equation given above.

Figure 4.5 shows the results of rough calculations for determining the absolute and specific thrusts of the A-4 engine at sea level and in space, operating on 80% weight concentration ethyl alcohol and liquid oxygen, with p_k = 20 atm abs, n = 1.16, and different values of p_v (depending on the different lengths of the chamber nozzle).

The curves of this graph show that:

1) shortening the nozzle relative to its normal terrestrial length leads to a decrease in the engine's absolute and specific thrusts in operation at ground level, in the rarefied layers of the atmosphere, and in space;

2) lengthening the nozzle relative to its normal terrestrial length leads to a decrease in the engine's absolute and specific thrusts at blastoff and to increasing them considerably as the engine operates in the rarefied layers of the atmosphere;

3) increasing the area of the nozzle outlet section by 30% and its length by 6% relative to the normal-length nozzle at ground level increases the absolute and specific thrusts of the engine in space by approximately 5%, and its internal efficiency by almost 10%.

The engine thrust vector must concur strictly with the axis of the missile's flight. In practice it is extremely difficult to accomplish this condition, since, in fact, an engine's thrust in relationship to the missile is always somewhat noncoaxial.

Calculations show that disruption of the centrality of the thrust vector of a ZhRD by 0.5% can cause a lateral force of about 0.1% of the thrust of the engine. For a missile with a mass ratio of 7:1 at the end of the power phase, the lateral acceleration may reach 0.1 g, and the lateral velocity about 60 m/sec.

This circumstance compels the application of some means for steering the missile during blastoff and in flight, as well as for changing the direction of its flight if necessary. One may use gas rudders, which are started by a hydraulic system (steering engines) for this purpose in missiles of large and medium thrust. Gas rudders may cause a lateral force of up to 20% of the axial value of the engine's thrust.

If gas rudders are placed in the gas flow at the rear of the engine nozzle, the engine thrust decreases by the magnitude ΔP_{rul}* kg. The magnitude of the resistance of the gas rudders of a missile depends on their design, dimensions, turning angle in the gas flow, and their burnout level during the engine's operation. In ballistic missiles a turn by gas rudders is possible within the limits of $\pm 25°$, and their working turns are usually accomplished in the interval of $\pm 12°$. This shows that the forces operating on gas rudders in the missile's guided-flight phase change within wide limits.

The frontal drag for the A-4 long-range missile's gas rudders, in a nondivergent position, is equal to about 640 kg, and in a divergent position attains 1560 kg.

In calculations, thrust losses of liquid-fuel rocket engines due to gas rudders are usually accepted, in accordance with the statistical data, as equal to about 1-3% for finned missiles and to about 2-4% for finless missiles, computed from ground-level engine thrust. This is convenient from the practical point of view, since one may determine the resistance of the gas rudders experimentally at ground level.

The quantity ΔP_{rul} is practically independent of flight altitude, since density and velocity of the stream of gases flowing out of the engine nozzle do not depend on the pressure of the surrounding medium, i.e., are not functions of flight altitude.

Instead of gas rudders, in some cases special steering combustion chambers, which may also be used as the final stage of an engine, are applied for steering a missile in flight. By changing the corresponding mode of fuel flow rate to the separate chambers, one may turn the missile in the required direction.

---

*[$\Delta P_{pyл} = \Delta P_{rul} = \Delta P_{rul'} = \Delta P_{rudder}$.]

If the fuel system contains a turbopump unit whose exhaust gas flows into the atmosphere through the expanding nozzle of an exhaust pipe which is parallel to the engine axis, a supplementary engine thrust is developed:

$$\Delta P_{THA} = \varepsilon_1 G'_s P_{ys.TNA},$$

where ε_1 is the factor accounting for the shortage of specific thrust due to part of the fuel being expended in the gas generator (as a consequence of partial use of the gases' heat differential in the turbine and their partial ejection through the constant-pressure valve if such a valve exists); G'_s is the fuel flow rate to the gas generator of the turbopump unit in kg/sec; $P_{ud.TNA}$ is the specific thrust developed by the gases when they flow out of the exhaust pipe of the turbopump unit's nozzle in kg thrust/kg fuel/sec.

Experience shows that the value of ΔP_{TNA} may amount to about 0.2 to 0.5% of the thrust of the engine chamber.

Allowing for this, the summary absolute thrust of an engine is expressed by the equation

$$P_s = P_s + \Delta P_{THA} - \Delta P_{py.s} = \varepsilon_2 P_s + \Delta P_{THA} = \varepsilon_2 G_s P_{ys.s} + \varepsilon_1 G'_s P_{ys.TNA},$$

where ε_2 is the factor allowing for the engine's thrust loss as a consequence of gas rudders at the rear of the nozzle.

The absolute thrust of a ZhRD at ground or sea level is an extremely important characteristic of the engine, since it determines the stability of the weapon at the instant of launching.

The magnitude of engine thrust at ground level is usually dependent upon the weapon's tactical designation. Thus, for example, for existing long-range missiles of the A-4 type, the magnitude of the thrust at ground level must be almost twice as great as the missile's launching weight.

Requirements for increasing range and weight of payload of a weapon leads to a great increase in the engine's absolute thrust. Toward the end of the Second World War, engine thrusts reached 26 tons.

Experience has shown that the development of a single-chamber engine is not practicable for extremely great thrusts because of the excessive increase in the dimensions and weight of the engine and the difficulties in its experimental development. The published literature reports that single-chamber engines with a thrust of several hundred tons may be developed.

Probably missiles with greater thrusts will be multichambered with synchronized thrust of the separate combustion chambers. Such an engine permits steering the missile's flight by differential throttling of the separate combustion chambers or turning them relative to the missile axis.

In practice, the thrust of a multichambered engine may attain extremely great magnitudes.

Calculations show that an engine in the form of clusters of several chambers of smaller size has a smaller total weight than a single-chambered engine of the same thrust operating under the same conditions. The advantage of a multichamber engine also lies in the convenience of regulating thrust; however, its external dimensions are greater than the dimensions of a single-chamber engine.

One must bear in mind that a liquid-fuel rocket engine, at launching, usually works up to full thrust only after a short interval of time after the trigger pulse is delivered. Again, in cutting off (stopping) the engine, its thrust does not cease simultaneously, but, as they say, the phenomenon of aftereffects is observed, leading to range dispersion, especially in ballistic missiles.

The magnitude of the aftereffect momentum and some amount of dispersion of its values, in a ZhRD, depends upon the length of time that fuel which was not cut off, because of the design particulars of the engine, continues to arrive in the combustion chamber and burn there.

At the moment the cut-off signal is given for the fuel components, some quantity of fuel is arriving in the chamber, where it burns after a certain time interval equal to its delay in ebullition. The cut-off valves are late in operating after the signal is given for their closure; the process of closing also does not take place simultaneously, since at this moment a certain amount of fuel continues to arrive in the chamber and burn there. After the cut-off valves close, the fuel which was not cut off rushes into the combustion chamber partly as a consequence of inertial effect, and also by partial vapor formation because of the decrease in the pressure in the lines, especially in the cooling duct of the engine chamber. Besides this, at the moment the supply of the components is cut off, there is some amount of gases of a certain pressure in the chamber. Because of all these things, the engine has an aftereffect momentum, which leads to a considerable range dispersion in ballistic missiles.

In order to decrease the range dispersion of these missiles, sometimes, before complete cutoff of fuel delivery to the combustion chamber, the engine at first shifts to operation with a smaller per-second flow rate, i.e., at the final thrust stage. For this same purpose, the extremely perfected instruments in systems for guiding the missile in the power phase of its trajectory are applied. Decreasing the scattering (dispersion) of the range of long-range missiles is possible by means of decreasing the liquid-fueled rocket engine's aftereffect momentum, as well as by careful calculation of its magnitude in working out the mechanism for separating the warhead from the body of the missile.

SECTION 5. ENGINE CHAMBER NOZZLE OPERATION AT NONOPTIMUM REGIMES

At the rear of the chamber nozzle, when operating on nonoptimum regimes, transformation of the supersonic gas flow into a subsonic gas flow occurs, with the restoration of the pressure from p_v to p_a, because of the system of compression waves (Fig. 4.6).

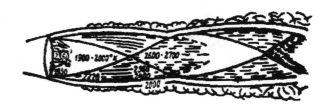

Fig. 4.6. Transformation of gas flow in the rear of an engine nozzle by compression waves as the consequence of underexpansion, relative to the pressure of the atmospheric air (when $p_v > p_a$), of the exhaust from the nozzle.

In the ranges of the operating regimes for underexpansion and overexpansion of the gases in the engine nozzle exhaust section, ordinarily the rated exhaust velocity w_v is established, whose magnitude is indepedent of the conditions of the surrounding medium and is determined only by the parameters of the gases in the combustion chamber and the geometrical form of the nozzle.

In cases where the nozzle operates with $p_v > p_a$ the stream of gas at the end of the nozzle at first expands with an increase in velocity, and then, as a consequence of its inertial velocity of overexpansion, is subjected to compression by the influence of the relatively great pressure of the surrounding layers of air on it, and so forth. These processes of overexpansion and overcompression of the gas stream and the layers of the atmospheric air adjoining it, relative to its pressure in an unperturbed condition, are usually accompanied by oblique compression wave.

The further a nonoptimum operating regime of the chamber nozzle varies from optimum, the stronger the perturbation in the stream of gases at the end of the engine chamber.

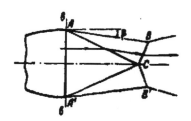

Fig. 4.7. Transformation of gas flow at rear of engine nozzle through compression waves as a consequence of its overexpansion relative to the atmospheric air pressure (when $p_v < p_a$).

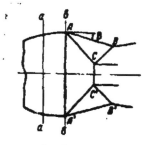

Fig. 4.8. Transformation of flow of exhaust gases from an engine nozzle with appearance of a normal compression wave due to the considerable overexpansion of gases relative to the atmospheric air pressure.

When a nozzle operates with $p_v < p_a$, transformation of the supersonic gas flow to subsonic is considerably more complex (Figs. 4.7 and 4.8). Here the pressure of the gas is restored to the pressure of the surrounding medium by means of a system of oblique and even, possibly, normal compression waves.

A decrease in the pressure ratio p_v/p_a causes an increase in the angle β (see Fig. 4.7), and when it exceeds a certain limiting value β', depending on Mach number (M_v), the gas-flow transformation scheme in the rear of the nozzle is even more complicated (see Fig. 4.8). In this case, a normal (or, more exactly, curvilinear) compression wave appears around the gas-flow axis, and this, when p_v/p_a is decreased to a fixed value, will be shifted to the nozzle's outlet section.

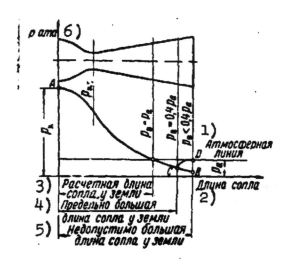

Fig. 4.9. Compression waves in flow in an engine nozzle as a consequence of excessive over-expansion relative to atmospheric air pressure. 1) Atmosphere line; 2) nozzle length; 3) rated nozzle length at ground level; 4) maximum nozzle length at ground level; 5) inadmissible nozzle length at ground level; 6) p, atm abs.

Until this normal compression wave approaches the chamber-nozzle outlet section, exhaust velocity will be supersonic, since in this case its magnitude is determined only by the parameters of the gases at the end of the combustion chamber and the design of the nozzle. By this means, the pressure in the gas flow at the rear of the nozzle will be restored up to the pressure of the atmospheric air by means of oblique compression waves.

After the normal compression wave moves to the chamber-nozzle exhaust section, gas outflow velocity will be sonic.

With a further decrease in p_v/p_a, the gases' normal compression wave enters the nozzle. Then, in the nozzle behind the plane of this compression wave, the gas stream becomes subsonic and strongly perturbed, as a result of which it may become detached from the surface of the nozzle wall (Fig. 4.9), and to considerable decrease of the engine's specific thrust.

In practice, we find that the gases' detachment from the wall of the chamber nozzle in a liquid-fueled rocket engine may occur when $p_v/p_a = 0.3$-0.4.

Regimes when normal compression waves occur within the nozzle are inefficient, and are therefore not desirable during normal operation of the engine. Such regimes may take place only when starting and stopping the engine, and also during test-stand trials of high-altitude engines.

For a variation in a <u>ZhRD</u> chamber operating regime until normal compression waves occur in the gases inside the nozzle, engine thrust may be determined by the above-mentioned general equations (4.8) and (4.9); in other cases these equations will give incorrect results to the computations.

Studies have established that the gas streams behind the nozzle of an engine affect the aerodynamic characteristics of the missile by decreasing frontal drag at supersonic flight speeds and increasing it at subsonic flight speeds.

With a subsonic flight speed the gas streams flowing out of the engine nozzle with a supersonic velocity operate like an ejector and create an inflow of the surrounding air behind the flying missile. This phenomenon considerably affects such devices as missiles, aerial torpedoes, and some types of jet aircraft, in which the engine is installed in the conical tail section.

The ejecting action of the gases flowing out of the engine nozzle creates an acceleration of the surrounding air and, consequently, a local increase in surface friction, as well as a decrease in the pressure on the surface of the missile close to the nozzle. This leads to an increase of the frontal drag through the moving air and a de-

- - -

crease in the atmospheric pressure at the exhaust from the nozzle (this decrease in pressure causes an increase in thrust, which is developed by the difference of pressures between the gases at the exhaust from the nozzle and the atmospheric air).

During the flight of a missile with supersonic speed, a turbulent wake with decreased local pressure is excited at its tail section. Under the influence of the reactive stream, the area of lowered pressure is filled by the gases flowing out of the engine, which leads to an increase in the pressure on the missile's tail section. By this means frontal d r a g is decreased, i.e., thrust is increased.

SECTION 6. THE THRUST COEFFICIENT OF AN ENGINE

To simplify calculations, the concept of an engine's thrust coefficient, which is the ratio of the chamber's absolute thrust P to the product of the gas pressures p_k in the combustion chamber and the area F_{kr} of the nozzle's critical section, i.e.,

$$K = \frac{P}{p_k F_{kr}},$$ (4.13)

has been introduced into liquid-fueled rocket engine theory.

This coefficient shows how many times the thrust of the engine chamber is increased relative to the basic term of thrust $(p_k F_{kr})$ by means of the convergent and divergent parts of the nozzle.

Using the concept of an engine chamber's thrust coefficient, the general thrust equation

$$P_a = \frac{Q_i}{g} w_a + F_a p_a - F_a p_1 = P_a - F_a p_1$$

may be given the following dimensionless form:

$$\frac{P_a}{p_k F_{kr}} = \frac{P_a}{p_k F_{kr}} - \frac{F_a p_a}{p_k F_{kr}} *$$

_____

*[P_H = $P_{absolute}$; P_π = P_{space};
F_p = F_r = $F_{raschetnyy}$ = F_{rated} (calculated).]

124

or

$$K_{\text{в}} = K_{\text{п}} - f_{\text{в}} \frac{p_{\text{а}}}{p_{\text{к}}},$$ (4.14)

where $K_n = p_n/p_k F_{kr}$ is the thrust coefficient of an engine when the gases flow out of the nozzle into the atmosphere; $K_p = P_p/p_k F_{kr}$ is the thrust coefficient of an engine when the gases flow out of the nozzle into space; and $f_v = F_v/F_{kr}$ is the relative area of the nozzle's exhaust section.

By this means, we will obtain the following additional expressions for determining the thrust of an engine chamber:

$$P = p_{\text{к}} F_{\text{кр}} K; \quad P_{\text{п}} = p_{\text{к}} F_{\text{кр}} K_{\text{п}}; \quad P_{\text{в}} = p_{\text{к}} F_{\text{кр}} K_{\text{в}}.$$

Bearing in mind that the absolute thrust of a chamber is also expressed by the equation

$$P_{\text{в}} = p_{\text{к}} F_{\text{кр}} \sqrt{\frac{2n^2}{n-1} \left(\frac{2}{n+1}\right)^{\frac{n+1}{n-1}} \left[1 - \left(\frac{p_{\text{в}}}{p_{\text{к}}}\right)^{\frac{n-1}{n}}\right]} + $$
$$+ F_{\text{в}}(p_{\text{в}} - p_{\text{а}}) = p_{\text{к}} F_{\text{кр}} K_{\text{в}},$$

we will obtain

$$K_{\text{в}} = \sqrt{\frac{2n^2}{n-1} \left(\frac{2}{n+1}\right)^{\frac{n+1}{n-1}} \left[1 - \left(\frac{p_{\text{в}}}{p_{\text{к}}}\right)^{\frac{n-1}{n}}\right]} + \frac{F_{\text{в}}(p_{\text{в}} - p_{\text{а}})}{p_{\text{к}} F_{\text{кр}}}.$$ (4.15)

This formula shows that K_n is also dependent upon atmospheric air pressure p_a, the value of which, with an increase in altitude, is lowered, and in space is equal to zero. Therefore, at a given operating regime in the chamber, the thrust coefficient K_n is increased with a rise in altitude, and in space attains maximum value, K_p.

Since the thrust of the chamber in space is expressed by the equation $P_p = F_v p_v (1 + nM_v^2)$, the thrust coefficient of a liquid-fueled rocket engine, when operating in space, may be expressed as follows:

$$K_{\text{в}} = \frac{P_{\text{в}}}{p_{\text{к}} F_{\text{кр}}} = \frac{F_{\text{в}} p_{\text{в}} (1 + nM_{\text{в}}^2)}{p_{\text{к}} F_{\text{кр}}} = f_{\text{в}} \frac{p_{\text{в}}}{p_{\text{к}}} (1 + nM_{\text{в}}^2).$$ (4.16)

If, in this formula, we replace f_v and p_v/p_k with the already well-known expressions of them which are functions of Mach number (M_v), and then replace this figure with its value dependent on the velocity coefficient λ_v, or express the latter by the ratio p_v/p_k, after the appropriate transformations, we will obtain the following expressions for the coefficient of the thrust of an engine when operating in space:

$$K_s = \left(\frac{2}{n+1}\right)^{\frac{n+1}{2(n-1)}} \frac{1+nM_s^2}{M_s\sqrt{1+\frac{n-1}{2}M_s^2}} = \left(\frac{2}{n+1}\right)^{\frac{1}{n-1}} \frac{\lambda_s^2+1}{\lambda_s} =$$

$$= 2\left(\frac{2}{n+1}\right)^{\frac{1}{n-1}} \frac{n}{\sqrt{n^2-1}} \sqrt{1-\left(\frac{p_s}{p_k}\right)^{\frac{n-1}{n}}} \left[1+\frac{n-1}{2n}\frac{(p_s/p_k)^{\frac{n-1}{n}}}{1-(p_s/p_k)^{\frac{n-1}{n}}}\right]. \quad (4.16')$$

In accordance with these same formulas, one may also determine the theoretical engine thrust coefficient $K_{p.t}$* when operating in space if we replace the corresponding actual parameters with their theoretical values which were obtained during thermodynamic computation of the engine chamber.

The following relationship exists between the actual thrust coefficient K_p of the engine when operating in space and its theoretical value $K_{p.t}$:

$$K_s = \varphi_c K_{s.t} \quad (4.17)$$

where φ_s** is the correction factor for the nozzle efficiency of this specific engine chamber nozzle.

It is simplest to determine the theoretical coefficient of thrust in space by means of Eq. (4.16'), having replaced the polytropic exponent n in it with the specific heat ratio k.

In existing engines $K_p \approx 1.5$-1.8. By increasing p_k and decreasing n or k, the value of K_p is increased.

By considering these factors, one may also express an engine cham-

---

*[$K_{n.\tau}$ = $K_{p.t}$ = $K_{pustota.teoreticheskiy}$ = $K_{space.theoretical}$.]
**[φ_c = φ_s = φ_{soplo} = φ_{nozzle}.]

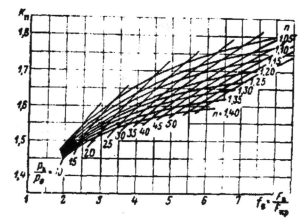

Fig. 4.10. Nomogram for determining geometrical dimensions of an engine nozzle.

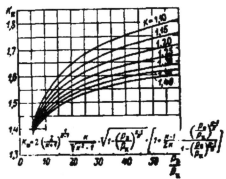

Fig. 4.11. Thrust coefficient of an engine in space, dependent on gas-expansion ratio in the nozzle and specific heat ratio k.

ber's thrust thus:

$$P_{\text{к}} = p_{\text{к}}F_{\text{кр}}\varphi_c K_{\text{к.т}} - F_{\text{в}}p_{\text{в}}; \quad P_{\text{в}} = p_{\text{к}}F_{\text{кр}}\varphi_c K_{\text{к.т}}$$

The formulas given here permit the determination of the basic geometrical dimensions of the chamber nozzle and some of its operating parameters. By this means, one may use the curves in Figs. 3.9, 4.10, and 4.11 for approximate calculations.

Example 1. Determine the absolute thrust of an engine when operating in space and the geometrical dimensions of the chamber nozzle, if $p_k = 40$ atm abs, $p_v = 1$ atm abs, $n = 1.15$, and $F_{kr} = 39$ cm^2.

Solution.

1. According to the graph in Fig. 4.10, when $p_k/p_v = 40/1 = 40$ and $n = 1.15$, we find $K_p = 1.725$ and $f_v = 6.5$.

2. Thrust of the engine chamber in space

$$P_{\text{в}} = p_{\text{к}}F_{\text{кр}}K_{\text{в}} = 40 \cdot 39 \cdot 1{,}725 = 2667 \text{ kg}.$$

3. Area of the nozzle cross section at outlet

127

$$F_{\text{в}} = f_{\text{в}} F_{\text{кр}} = 6,5 \cdot 39 = 253,5 \text{ см}^2.$$

4. Diameters of the critical and exhaust sections of the nozzle:

$$d_{\text{кр}} = \sqrt{\frac{4 F_{\text{кр}}}{\pi}} = \sqrt{\frac{4 \cdot 39}{3,14}} = 7,042 \text{ см}; \quad d_{\text{в}} = \sqrt{\frac{4 \cdot 253,5}{3,14}} = 17,92 \text{ см}.$$

We will also determine the thrust of the engine chamber in space and the pressure of the gases in the nozzle exhaust section when $p_k = 40$ atm abs, $n = 1.15$, $d_{kr} = 7$ cm, and $d_v = 21$ cm.

We solve this problem in the following manner.

1. The relative area of the exhaust section of the chamber nozzle will be

$$f_{\text{в}} = F_{\text{в}}/F_{\text{кр}} = d_{\text{в}}^2/d_{\text{кр}}^2 = 21^2/7^2 = 8,9.$$

2. According to the graph in Fig. 4.10, when $f_v = 8.9$ and $n = 1.15$, we find the level of gas expansion in the nozzle:

$$p_k/p_v = 60,$$

from whence $p_v = p_k/60 = 40/60 \approx 0.67$ atm abs.

3. According to the graph in Fig. 4.11, when $p_k/p_v = 60$ and $n = 1.15$, we find the chamber's thrust coefficient K_p in space is equal to 1.76.

4. We will determine the area of the critical section of the nozzle according to the formula

$$F_{\text{кр}} = \frac{\pi d_{\text{кр}}^2}{4} = \frac{3,14 \cdot 7^2}{4} = 38,48 \text{ см}^2.$$

5. The thrust of the engine chamber when operating in space:

$$P_{\text{в}} = K_{\text{в}} p_{\text{к}} F_{\text{кр}} = 1,76 \cdot 40 \cdot 38,4 = 2703,36 \text{ kg}.$$

SECTION 7. SPECIFIC THRUST OF AN ENGINE

The usual accepted evaluation of the operating economy of a ZhRD on a given fuel and on the corresponding regime is by the magnitude of specific thrust.

In calculating the characteristics of a liquid-fueled rocket en-

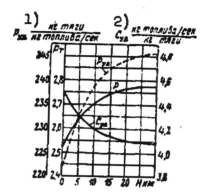

Fig. 4.12. Rated dependence of absolute thrust, specific thrust, and specific consumption on flight altitude. 1) P_{ud}, kg thrust/(kg fuel/sec); 2) C_{ud}, kg fuel/sec per kg thrust.

gine for evaluation of the fuel used in it and the quality of the operating cycle, the magnitude of the specific thrust, depending on the regime and operating conditions of the engine, should be determined by taking into consideration the per-second fuel-component flow rate to the combustion chamber and the flow rates for servicing the fuel-feed system and forming the protective curtain around the surface of the engine-chamber burner liner.

Calculations show that the specific thrust of an engine is quite dependent on flight altitude H of the weapon (see Fig. 4.12); the less pressure p_k in the combustion chamber, the greater the dependence of P_{ud} on H.

Considering the above, one should distinguish the specific thrust of the chamber only and of the engine as a whole, i.e.,

1. The engine chamber:

a) in general

$$P_{уд.к} = P_к/G_i;$$ (4.18)

b) when the nozzle is operating in the optimum regime ($p_v = p_a$),

$$P_{уд.опт} = P_{опт}/G_s = w_a/g;$$

c) when the nozzle operates in a nonoptimum regime ($p_v \gtrless p_a$),

$$P_{уд.н} = P_н/G_s = w_{эф}/g;\ *$$

d) when operating at ground or sea level (H = 0)

$$P_{уд.о} = P_о/G_i;\ **$$

$*[P_{уд.н} = P_{ud.n} = P_{udel'nyy.neoptimal'nyy} = P_{specific.nonoptimum}.]$
$**[P_{уд.о} = P_{ud.0} = P_{udel'nyy.0} = P_{specific.(H-)\ zero}.]$

129

e) when operating in space ($p_a = 0$), $P_{ud.p} = P_p/G_s$.

2. The engine as a whole:

a) in general

$$P_{уд\,\Sigma} = P_\Sigma/G_\Sigma; \qquad\qquad (4.19)$$

b) when the nozzle is operating in a nonoptimum regime ($p_v \gtrless p_a$),

$$P_{уд\,\Sigma\,в} = P_{\Sigma\,в}/G_\Sigma;$$

c) when operating at ground or sea level (H = 0)

$$P_{уд\,\Sigma\,0} = P_{\Sigma0}/G_\Sigma;$$

d) when operating in space ($p_a = 0$), $P_{ud\Sigma p} = P_{\Sigma p}/G_\Sigma$.

The specific thrust of an engine chamber when operating in space may also be expressed by the formula

$$P_{уд.\,п} = P_п/G_s = p_к F_{кр} K_п/G_s = \beta K_п$$

where $\beta = p_k F_{kr}/G_s$ is the actual gas-pressure impulse in the combustion chamber in kg-sec/kg.

Thus, the specific thrust of an engine chamber, when the nozzle is operating in a nonoptimum regime ($p_v \neq p_a$), will be

$$P_{уд.в} = \frac{P_в}{G_s} = \frac{P_п - F_в p_в}{G_s} = \beta K_п - f_{уд.в} p_в =$$

$$= \beta K_п - \beta f_в \frac{p_в}{p_к} = \beta \left(K_п - f_в \frac{p_в}{p_к} \right).$$

where $f_{ud.v} = F_v/G_s$* is the true specific area of the exhaust section of the chamber nozzle in kg/cm^2; and $f_v = F_v/F_{kr}$ is the relative area of the exhaust section of the chamber nozzle.

In liquid-fueled rocket engine theory, there is also an accepted distinction of the theoretical specific thrust of an engine chamber:

$$P_{уд.т} = w_{в.т}/g.\,**$$

Since the actual operating characteristics of a chamber may be

---

*[$f_{уд.в}$ = $f_{ud.v}$ = $f_{udel'nyy.vykhod}$ = $f_{specific.outlet}$·]
**[$P_{уд.т}$ = $P_{specific.theoretical}$·]

expressed in the following manner by means of the corresponding theoretical characteristics computed during the thermodynamic computation of the engine:

$$K_{\text{в}} = \varphi_c K_{\text{в.т}}; \quad \beta = \varphi_{p_{\text{к}}} \beta_{\text{т}}; \quad P_{\text{уд}} = \varphi_{p_{\text{д}}} P_{\text{уд.т}}; \quad f_{\text{уд.в}} = \varphi_{p_{\text{к}}} f_{\text{уд.в.т}} \ast$$

the specific thrust of the engine chamber may be determined thus:

$$P_{\text{уд.в}} = \varphi_{p_{\text{к}}} \beta_{\text{т}} \varphi_c K_{\text{в.т}};$$

$$P_{\text{уд.д}} = (\beta_{\text{т}} \varphi_c K_{\text{в.т}} - f_{\text{уд.в.т}} p_{\text{в}}) = \varphi_{p_{\text{к}}} \beta_{\text{т}} \left(\varphi_c K_{\text{в.т}} - f_{\text{в}} \frac{p_{\text{в}}}{p_{\text{к}}} \right);$$

$$P_{\text{уд.в}} = \varphi_{p_{\text{д}}} P_{\text{уд.т}} + \varphi_{p_{\text{к}}} f_{\text{уд.в.т}} (p_{\text{в}} - p_{\text{н}}),$$

where $\varphi_{p_{\text{к}}} = \beta / \beta_t$ is the gas-pressure coefficient (correction factor) for the engine combustion chamber; and $P_{\text{ud}} = w_v / g$ and $P_{\text{ud.t}} = w_{v.t} / g$ are the true and theoretical dynamic specific thrusts, respectively, of the engine chamber.

The chamber's specific thrust $P_{\text{ud.n}}$, during nonoptimum nozzle operation, may be converted into the specific thrust $P_{\text{ud.opt}}$ during optimum nozzle operation according to the formula

$$P_{\text{уд.опт}} \approx P_{\text{уд.д}} \frac{\sqrt{1 - (p_{\text{в}}/p_{\text{к}})^{\frac{n-1}{n}}}}{\sqrt{1 - (p_{\text{в}}/p_{\text{н}})^{\frac{n-1}{n}}}} \frac{\text{kg thrust}}{\text{kg-fuel/sec}}. \tag{4.19'}$$

In accordance with an analogous formula, one may evaluate the specific thrusts of engines operating at various values of p_k and the same values of p_v. The magnitude of specific thrust is affected by the regime and operating conditions of the engine, its design particulars, and other factors, such as:

1) type of fuel, its composition, and method of atomization;

2) pressure in the combustion chamber;

3) level of gas expansion in the nozzle;

4) shape and dimensions of the combustion chamber;

5) engine-nozzle divergence angle and configuration;

$\ast \ [\varphi_{p_k} = \varphi_{p_{\text{chamber}}} .]$

6) relative area ($f_k = F_k/F_{kr}$) of the combustion chamber;

7) engine cooling system;

8) nature of the methods used to protect the chamber liner and nozzle liner from overheating;

9) propellant-feed system, its design and performance, and operating economy;

10) pressure at which fuel is fed to the combustion chamber;

11) flight altitude (see Fig. 4.12), etc.

By the improvement of vaporization and mixing of fuel components, and by eliminating enrichment of the peripheral spray with combustible, which is often used to protect the chamber liner from overheating, as well as using a number of other methods, one may considerably increase an engine's specific thrust.

When raising the pressure in the combustion chamber to $p_k \approx 60$ atm abs, at first the specific thrust increases sharply, from 60 to 100 atm abs its growth gradually slows down, and at about 200 atm abs it remains almost constant (Figs. 4.13 and 4.14).

Increasing P_{ud} by raising p_k above a certain value is limited in practice. Raising p_k, other things being equal, causes:

1) decrease in the external dimensions of the combustion chamber because of the smaller specific volume of the fuel combustion products formed;

2) increase in the thermal efficiency η_t of the engine because of the increase in the gas-pressure difference in the nozzle;

3) increasing the heat liberation coefficient φ_k of the fuel in the combustion chamber as a consequence of the intensification in the working cycle by decreasing the dissociation of the gases and increasing the physical coefficient $\varphi_{p.k}$* of completeness of fuel combustion,

*[$\varphi_{\text{п.к}} = \varphi_{p.k} = \varphi_{\text{polnota.kamera}} = \varphi_{\text{completeness (of combustion).chamber}}$]

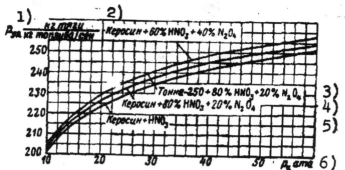

Fig. 4.13. Dependence of an engine's theoretical specific thrust on combustion chamber pressure for several fuels when p_v = 1 atm abs and α = 0.8.
1) P_{ud}, kg thrust/(kg fuel/sec); 2) kerosene + 60% HNO_3 + 40% N_2O_4; 3) Tonka-250 + 80% HNO_3 + 20% N_2O_4; 4) kerosene + 80% HNO_3 + 20% N_2O_4; 5) kerosene + HNO_3; 6) p_k, atm abs.

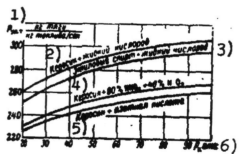

Fig. 4.14. Theoretical specific engine thrust as dependent on combustion chamber pressure for several fuels when p_v = 1 atm abs and α = 0.8. 1) $P_{ud.t}$, kg thrust per kg fuel/sec; 2) kerosene and liquid oxygen; 3) ethyl alcohol and liquid oxygen; 4) kerosene + 60% HNO_3 + 40% N_2O_4; 5) kerosene + nitric acid; 6) p_k, atm abs.

which, in conjunction with the first factor, increases the velocity of the gases in the nozzle exhaust and, proportionally, the specific thrust of the engine;

4) decreasing the time τ_{pr}* the fuel remains in the combustion chamber, which also decreases its required volume, dimensions, and weight.

However, beginning at some magnitude of p_k, increasing it further does not give noticeable advantages.

On the other hand, raising p_k basically leads to the following phenomena:

---

*[τ_{np} = τ_{pr} = $\tau_{prebyvaniye}$ = τ_{stay}.]

1) raising the temperature in the engine combustion chamber and complication of its cooling system;

2) changing the engine-chamber nozzle length due to the necessity for setting up a great gas-pressure difference across the nozzle;

3) increasing the weight of the engine as the consequence of the requirement for greater strength of the combustion chamber, nozzle, and other elements of the engine;

4) increasing the energy losses in supplying fuel to the combustion chamber under a great pressure head, and

5) increasing the cost of the engine.

In practice, gas pressure in the engine combustion chamber is chosen in each separate case on the basis of comparative calculations showing its effect on specific thrust, dimensions and weight of the chamber and propellant-feed system of the engine, on the possibility of effective cooling of the engine, etc.

In future engines, one may expect an increase in the ratio p_k/p_v by means of increasing p_k and decreasing p_v, which is especially suitable for long-range missiles and some types of aircraft.*

The design of reaction engines with a combustion chamber pressure of up to 100-200 atm abs and above is entirely possible, when atomic reactors, intended for using the heat of nuclear reactions, are constructed.

The most radical method of increasing P_{ud} is the use of exotic fuels. However, this means of increasing P_{ud} leads to an increase in T_k, which considerably complicates the problem of cooling the engine.

Calculations show that, in practice, the possible limit of increasing P_{ud} by means of the chemical energy of the fuel is about 350 to 450 kg-thrust/(kg-fuel/sec).

Magnitude of specific thrust, characterizing the engine's operating economy, does not reveal the specific nature of the heat process occurring in the combustion chamber and in the engine nozzle, or the cycle's reserves for increasing operating economy. In this relationship, the engine efficiencies are the best indices.

Ground-level data for specific thrust have considerable value only for low-altitude engines.

The specific thrust in space should be considered the basic characteristic of the operating economy of high-altitude ZhRD.

The most complete characteristic of the operating economy for a high-altitude engine is its mean specific thrust with regard to the trajectory of the flight, expressed by the formula

$$P_{уд.ср} = \frac{\frac{a_g}{g}w_s + F_a(p_a - p_{a.cp})}{G_s} \frac{\text{kg thrust}}{\text{kg fuel/sec}},^* \qquad (4.19'')$$

where $p_{a.sr}$ is the mean atmospheric pressure in atm abs along the trajectory of the flight.

SECTION 8. FUEL FLOW RATES IN AN ENGINE

To determine the dimensions of a combustion chamber, chamber burner cup, fuel tanks, and propellant-feed system, also including start-control devices, one must know the per-second flow rate of the fuel components in the engine into the chamber and for supplying the feed system.

In a liquid-propellant rocket engine, in general, we accept the distinction of the following per-second fuel flow rates:

_____

$*[P_{уд.cp} = P_{ud.sr} = P_{udel'nyy.srednyaya} = P_{specific.mean};$
$P_{a.cp} = P_{a.sr} = P_{atmosfera.srednyaya} = P_{atmosphere.mean}.]$

1) <u>basic</u> — in the engine combustion chamber when it is operating in a given regime G_s in kg/sec;

2) <u>starting</u> — in the combustion chamber, corresponding to the operating regime of the engine at the first starting stage, G_s p^* in kg/sec;

3) <u>auxiliary</u> — in servicing the propellant-feed system (TNA, ZhAD) G'_s in kg/sec.

The per-second propellant flow rate to the combustion chamber, when the engine is operating on the corresponding operating regime, is determined in accordance with the formula

$$G_s = \frac{P_s'}{P_{уд}} = \frac{P_s}{w_{эф}/g} = \frac{gP_s}{w_{эф}} \text{ kg/sec.} \tag{4.20}$$

In planning and designing an engine, it is necessary to know the relationships between the per-second propellant-flow rate G_s into the chamber, the magnitude of the area F_{kr} of the cross section of the nozzle throat and the parameters of the gases at the entrance to the nozzle from the combustion chamber.

Gasdynamics give the following interdependence between G_s, F_{kr}, and the parameters of the gas at the entrance into the nozzle of a ZhRD:

$$G_s = p_к F_{кр} \sqrt{\frac{ng}{R_к T_к}\left(\frac{2}{n+1}\right)^{\frac{n+1}{2(n-1)}}} = p_к B = F_{кр} C \text{ kg/sec}$$

or $$\tag{4.21}$$

$$G_s = \frac{p_к F_{кр} A}{\sqrt{R_к T_к}} = p_к F_{кр} \frac{A}{\sqrt{R_к T_к}} = \frac{p_к F_{кр}}{\beta},$$

where $A = \sqrt{2g\frac{n}{n+1}\left(\frac{2}{n+1}\right)^{\frac{2}{n-1}}} = \sqrt{ng}\left(\frac{2}{n+1}\right)^{\frac{n+1}{2(n-1)}}$ — a coefficient which is only slightly dependent on polytropic exponent <u>n</u> (when <u>n</u> is increased from 1.1 to 1.2, the magnitude of A is increased from 1.98 to 2.03); and $\beta = \sqrt{R_k T_k}/A = p_k F_{kr}/G_s$, the specific impulse of the pressure of the

$\overline{*[G_{s\ п} = G_{s\ p} = G_s \text{ puskovoy} = G_s \text{ starting}\cdot]}$

136

gases in the combustion chamber, which characterizes the properties of the propellant and the completeness of its combustion.

In the case of decrease in gas pressure along the length of the combustion chamber we have

$$G_s = p_k' F_{\text{кр}} \sqrt{\frac{n g}{R_k' T_k} \left(\frac{1 + \frac{n-1}{2} M_k^2}{\frac{n+1}{2}} \right)^{\frac{n+1}{2(n-1)}}} \quad \text{kg/sec.} \tag{4.22}$$

These formulas show that under conditions of constant temperature and composition of the products of propellant combustion, the values of p_k and F_{kr} change in direct proportion to the change in G_s.

Using these formulas with the assumption mentioned, one may compute the dependence of G_s on p_k or F_{kr}, and of thrust P on G_s or p_k, and construct the graphs accordingly.

For engines of great thrust, great per-second propellant-flow rates are required, which is one of the basic difficulties in the field of development of such engines.

The per-second flow rates of the oxidizer and combustible are determined from the equation of the summary propellant-flow rate to the combustion chamber:

$$G_s = G_{s\ g} + G_{s\ o} \quad \text{kg/sec;*}$$

bearing in mind that $G_{s\ o} = X G_{s\ g}$, we obtain

$$G_s = G_{s r} + \chi\, G_{s r} = G_{s r}\, (1 + \chi). \tag{4.23}$$

whence

$$G_{s r} = \frac{G_s}{1 + \chi} \quad \Big\} \text{kg/sec;}$$

$$G_{s o} = \frac{\chi G_s}{1 + \chi} \quad \Big\} \text{kg/sec,} \tag{4.24}$$

where $\chi = G_{s\ o}/G_{s\ g}$, the true weight coefficient of the propellant

---

*[$G_{s\ r}$ = $G_{s\ g}$ = G_s goryuchyeye = G_s combustible;
$G_{s\ o}$ = $G_{s\ o}$ = G_s okislitel' = G_s oxidizer.]

compound.

The per-second flow rates of the oxidizer and combustible to the combustion chamber of the engine may also be determined in accordance with the formulas:

$$G_{s.o} = g_o G_s \quad \text{kg/sec;}$$
$$G_{s.r} = g_r G_s \quad \text{kg/sec,}$$
(4.25)

where g_o and g_g are the weight proportions of the propellant components in kg/kg.

Some engines of great thrusts intended for long-range missiles are made with two stages of thrust, the starting and operating stages. For example, the A-4 engine has a starting and final thrust of 8 tons and an operating thrust of 26 tons. Shifting the engine from the first stage to the second is carried out after it has worked up normally to the previous starting stage.

A starting stage of thrust is introduced for the purpose of starting the engine with a per-second propellant-flow rate to the combustion chamber which is less than the flow rate in the operating regime (resembling the A-4 engine), and, by the same means, avoiding possible engine explosion during ignition of the propellant mixture which has accumulated in the combustion chamber in case of a malfunction in igniting the propellant during starting.

In some existing engines for long-range missiles starting flow rate $G_{s\ p} \approx 20\text{-}30\%$* of G_s.

The per-second propellant-flow rate G'_s in servicing the fuel-feed system, when operating in a rated regime, amounts to about 1.5-3% of the per-second propellant-flow rate G_s to the engine combustion chamber.

When engine thrust is decreased by means of throttling the per-

*$[G_{s\,n} = G_{s\ p} = G_{s}$ puskovoy $= G_s$ starting.$]$

second propellant-flow rate to the combustion chamber, the percentage of the propellant-flow rate used to start the turbine of the pump assembly grows as a consequence of the disruption, by this means, of the optimum operating conditions of the gas generator and turbine of the propellant-feed system. For example, with an engine thrust equal to 20% of the rated value, the propellant-flow rate for starting the TNA may amount to 10% of G_s.

The operating economy of the engine is sometimes evaluated by the magnitude of the specific fuel consumption:

1) per kg of thrust, per hour

$$C_{уд} = \frac{G_s}{P_a} 3600 = \frac{1}{P_{уд.s}} 3600 = \frac{1}{w_a} \text{ kg-fuel/kg-thrust}, \qquad (4.26)$$

2) per thrust-hour

$$C_P = \frac{G_s}{N_P} 3600 = \frac{G_s 75}{P_a V} 3600 = C_{уд} \frac{75}{V} \text{kg/hp-hr}, \qquad (4.27)$$

where V is the missile's flight speed in m/sec.

The specific fuel consumption in a ZhRD depends to a considerable degree on the quality of the propellant itself.

The theoretical specific consumptions for some fuels in a ZhRD at an optimum ratio of components, $p_k = 40$ atm abs and $p_v = 1$ atm abs, have the following approximate values:

	kg/m-sec
1) kerosene + 98% nitric acid	4.35
2) Tonka-250 and 98% nitric acid	4.09
3) kerosene + 60% of 98% concentration HNO_3 + 40% N_2O_4	3.98
4) Tonka-250 + 60% of 98% HNO_3 + 40% N_2O_4	3.95
5) dimazine (DMG)* + 60% of 98% HNO_3 and 40% N_2O_4	3.82
6) 93.5% ethyl alcohol + liquid oxygen	3.68
7) kerosene + liquid oxygen	3.56
8) hydrazine + chlorine trifluoride	3.51
9) dimazine (DMG) + liquid oxygen	3.41
10) hydrazine + liquid oxygen	3.36

*[ДМГ = DMG = dimetil gidrazin = unsymmetrical dimethyl hydrazine — UDMH (dimazine).]

11) hydrazine + nitrogen trifluoride...............	3.24
12) kerosene + fluorine monoxide..................	3.00
13) ammonia + liquid fluorine.....................	2.92
14) hydrazine + liquid fluorine...................	2.90

The figures indicated show that fluorine engines have a relatively small specific fuel consumption which is explained by the higher H_u of the propellant based on fluorine.

By increasing flight altitude, specific fuel consumption is decreased in inverse proportion to $P_{ud.n}$, if the inertia load is ignored.

Specific fuel consumption when the engine is operating on a given regime in space is independent of flight altitude and is of a constant magnitude.

One may consider that a 1% savings in fuel consumption, other conditions being equal, would increase the payload of a missile by approximately 10%.

Example 2. An engine develops a thrust of 1000 kg at a per-second fuel-flow rate to the combustion chamber of 4.86 kg. Determine the specific fuel consumption in the engine.

Solution.

1) specific thrust of the engine

$$P_{ud} = P/G_s = 1000/4.86 = 206 \text{ kg-thrust/(kg-fuel/sec)};$$

2) specific fuel consumption

$$C_{ud} = 1/P_{ud} = 1/206 = 0.00485 \text{ (kg-fuel/sec)/kg-thrust} =$$
$$= 4.85 \text{ (kg-fuel/sec)/m-thrust.}$$

SECTION 9. METHODS AND LIMITS OF REGULATING ENGINE THRUST

In some cases in a ZhRD the requirement arises for the possibility of regulating the magnitude of thrust within greater or smaller limits. This may be required, for example, to accomplish a given rule for changing the thrust and acceleration of the rocket from time to time, obtaining a given flight speed at the instant the engine is cut off, or compensating the changes in the characteristics of the engine units and the change in the atmospheric pressure with altitude, to ensure an optimum operating regime, and by this means to obtain the best thrust characteristics.

A liquid-propellant rocket engine with an invariable per-second fuel-flow rate to the combustion chamber gives the missile a flight of changing acceleration, because of the decrease in the back pressure of the atmosphere with rise in altitude (increase of the static thrust component) and the decrease of the missile's weight as fuel components are consumed in the missile. In practice, for the flight of ballistic rockets in an atmosphere of fixed density, a flight with constant acceleration may be more practicable, which may be accomplished by regulating the engine thrust. During these rockets' flight in space, it is advantageous to have an engine with constant thrust. Calculations show that to obtain the best flight characteristics for rockets of various designations, it is necessary to regulate the engine thrust from time to time in accordance with a set of optimum rules. This permits a considerable increase of the payload of the rocket in comparison with its value during a flight with constant thrust.

In cases of application of ZhRD for piloted aircraft, the regulation of their thrust from time to time must be carried out in accordance with the most advantageous climb conditions, for providing maximum speed, range, and duration of flight. For example, to obtain the maximum flight speed of an aircraft, the greatest possible engine thrust is required, while for maximum range and duration of flight it is necessary for the engine to operate with a relatively small thrust. For an aircraft's flight in cruising regimes and at high altitudes, one may regulate the thrust in order to provide the optimum expansion of gases in the chamber nozzle.

In practice, regulating the thrust of a ZhRD may be accomplished by means of:

1) changing the per-second fuel-flow rate to the combustion chamber (throttle control of the engine), i.e., changing the pressure in it;

2) switching separate combustion chambers in or out of operation, if the engine is multichambered.

It is possible to change the per-second fuel-flow rate to the combustion chamber of an engine by means of:

a) changing the fuel-flow pressure to the combustion chamber through a constant number of injectors; for example, regulating the fuel feed to the gas generator, as a result of which the number of revolutions of the turbine and the capacity of the pumps of the fuel-feed system of the engine are changed;

b) adaptation of rotary valves (resembling the BMW-R3386 engine), by means of which part of the injector orifices in the combustion-chamber head are opened or closed as necessary for feeding fuel;

c) adaptation of a group-injector system for the same purpose (resembling the "Walther" 109-509A engine), permitting switching of separate groups of fuel injectors in or out of operation as convenient.

In the BMW-R3386 engine, turning the regulating valves is accomplished, depending on change of altitude and flight speed, by a special automatic device for keeping a constant Mach number. In the "Walther" 109-509A engine, the chamber has 12 injectors of the flat-spray type. All of these injectors are connected in two groups of three injectors each and one of six injectors. The pilot controls the operations of these groups of injectors as convenient.

The throttle control method of regulating the thrust of a ZhRD is the simplest, but it is extremely uneconomical, since by increase or decrease of the per-second fuel-flow rate to the combustion chamber in relationship to its optimum value, the following factors deteriorate:

1) the processes of vaporization and mixing of the fuel components, because of the change in the normal pressure difference of the components in the injectors;

2) the quality of the cycle in the engine chamber, as a consequence of a decrease in the pressure of the gases in it, deterioration in the quality of vaporization, and the fact that the volume of the combustion chamber and the dimensions of the nozzle do not correspond to the changed per-second fuel-flow rate;

3) the state of engine chamber cooling by the regenerative method, especially when one of the propellant components serves as the coolant; this may lead to overheating and even burning out the inner liner of the chamber as a consequence of the decreased amount of fuel because of the decreased thrust and, hence, deterioration in the conditions of heat transfer to the liner, which do not correspond to the changed thermal operating regime of the engine.

Besides this, the pressure in the combustion chamber is lowered by a decrease in the fuel-flow rate, and therefore the level of dissociation of the products of combustion is increased, which, in general, decreases the specific thrust of the engine and disrupts the stability of its operation.

This shows that regulating the thrust of an engine by throttle control may be practicable only within comparatively narrow limits.

Loss ΔP_{ud} of specific thrust by throttle control of the fuel-flow rate to the chamber of the engine depends on flight altitude and intensity of throttling. With an increase in altitude, the loss ΔP_{ud} decreases, and in space is theoretically equal to zero. One may consider that a five-fold decrease in the engine chamber's thrust by throttle control at sea level is accompanied by a decrease in the specific thrust of almost 20%, and losses in thrust of approximately 8% of nominal value.

Decreasing the loss in specific thrust, which is conditional upon the decrease of the pressure difference in the injectors because of throttle control of the fuel-flow rate, is possible by regulating the thrust by means of application of the above-mentioned rotary valves or by a group system of injectors. In these methods of regulation, those injectors which are not switched off operate at rated (optimum) pressure differences. However, such thrust regulation is also accompanied by an unavoidable deterioration in the method used in cooling the engine chamber, because of the reasons mentioned above. Besides this, switching off part of the injectors does not provide equal distribution of the fuel components across a cross section of the combustion chamber, the propellant combustion in the relatively large volume of the chamber deteriorates, and a fusion of the nonoperating part of the injectors is caused as the consequence of the inflow of gases of high temperature through them, as well as causing some asymmetry in the engine-chamber thrust.

Satisfactory propellant-component vaporization during regulation of engine thrust by throttle control of the fuel-flow rate is made possible by using injectors with regulated nozzles. However, such injectors are complicated in construction.

Multichambered engines permit stage regulation of thrust by cutting the separate chambers in or out of operation in the optimum regime without decreasing the operating economy of the engine. However, multichamber (bloc) engines have comparatively complicated fuel-feed systems because of the necessity of synchronizing the thrust of the separate chambers, and greater dimensions. In changing the per-second fuel-flow rate to the combustion chamber of engines operating with variable thrust, the pressure of the gases in the chamber must be constant; considerable decrease in its magnitude leads to deterioration of the cycle in the combustion chamber, to the origin of an unstable propellant-combustion regime, and other unfavorable consequences.

Consequently, regulating the thrust of a ZhRD by means of throttle control of the per-second fuel-flow rate is accompanied by a decrease in the operating economy of the engine, disruption of the normal conditions of chamber cooling, and the origination of unstable operation.

The lower limit for reducing the thrust of an engine by throttle control of the fuel-flow rate may, in practice, be established either by the given coolant's maximum permissible heat in the chamber passage, or by compression waves originating in the gases in the nozzle, which may take place when the ratio of the pressures of the gases in the outlet from the nozzle and in the atmospheric air is about $p_v/p_a \leq 0.4$.

The ratio of pressures mentioned above shows that, in practice, the possible intensity of throttling an engine for changing the thrust, according to this parameter, is less at ground level and at low altitudes than at high altitudes. One may consider that by lowering the thrust and the pressure of the gases in the combustion chamber of the engine by 80%, the temperature of the coolant may be raised by almost 50%.

If the coolant permits only an insignificant rise of temperature, or if an extremely intense throttling of the engine chamber is required for decreasing the thrust, then it may become necessary to change the composition of the propellant for decreasing the temperature of the products of propellant combustion at extremely small thrusts, to use a film protection against overheating of the chamber liner, and so forth.

Regulating the thrust of an engine below 10% of its nominal value is, in practice, impossible to accomplish because of the instability of the combustion of the propellant in the combustion chamber, excessive increase in the temperature of the coolant, and the sharp decrease in specific thrust. In some existing types of aircraft ZhRD which operate on nitric acid and kerosene, the thrust is regulated within the limits of 30-100%. Somewhat wider limits of thrust regulation (12-100%) are possible for engines operating on hypergolic propellant components (Table 4.1).

A good start-up and smooth thrust control of an engine within wide limits may be provided only by preserving the necessary propellant-component ratio. This applies also to the composition of the fuel used to operate the gas generator. If these conditions are not fulfilled, an excessively high temperature in the chamber of the gas generator or uneven combustion may take place, and the reaction may even stop.

In designing any system for regulating an engine, it is necessary to consider the requirements demanded of the system during regimes in launching, normal operation, and operation at the instant of propellant cutoff. In general, very precise regulation of all three operating regimes of the engine is desirable; however, the designer's potentialities in this relationship are limited for reasons of a practical nature.

This circumstance leads to the emergence of new problems, particularly in providing pumps of identical characteristics, and to the complication of the system of regulation (the presence of several feed-

TABLE 4.1

Limits of Thrust Regulation of Some Existing
ZhRD

1) Марка двигателя	2) Назначение двигателя	3) Способ регулрования тяги	4) Тяга двигателя кг		7) Пределы регулрования тяги %
			максим. 5)	миним. 6)	
8) „Вальтер" 109-509А	9) Авиационный	10) Групповые форсунки	1700	200	11,7
11) БМВ-Р3386	12) Зенитный снаряд	13) Поворотные золотники	380	60	15,7
14) „Скример"	15) Авиационный	16) Дросселирование расхода	3600	450	12,5

1) Brand of engine; 2) purpose of engine; 3)
method of thrust regulation; 4) engine thrust,
kg; 5) maximum; 6) minimum; 7) limits of
thrust regulation, %; 8) "Walther" 109-509A;
9) aircraft; 10) group injectors; 11)
BMW-R3386; 12) antiaircraft missile; 13) rotary
valves; 14) "Screamer"; 15) aircraft; 16) flow-
rate throttling.

back circuits). The use of a bloc of single-chambered engines leads to
further complication of the system as the result of the emergence of
several regulation circuits.

Since ZhRD with thrust regulated in accordance with a fixed
rule are very complicated in structure, their application has turned
out to be impractical in a number of cases. For this reason the thrust
of the majority of the contemporary missiles which are being manufactured is not regulated. The engines of long-range missiles may accomplish thrust regulation in the form of one stage during launching and
several seconds before engine cutoff, which creates the necessity of
providing reliable starting and stopping of the engine according to a
given program.

SECTION 10. FLOW-RATE CHARACTERISTICS OF AN ENGINE

The dependence of absolute and specific thrusts (P and P_{ud}) of an engine, in the presence of fixed values of altitude and flight speed, on the pressure p_k in the combustion chamber, i.e.,

$$P \text{ and } P_{ud} = f(p_k),$$

is called the flow-rate or throttle characteristics of a ZhRD; these characteristics may also be dependent upon the factors determining the magnitude of its pressure p_k, i.e., per-second fuel-flow rate C_s [sic] to the combustion chamber; feed pressure p_p of the propellant components to the combustion chamber; and number of revolutions $\underline{n}$ of the turbo-pump unit, if one is included in the engine fuel-feed system, and other factors.

By means of the flow-rate characteristics, the most advantageous operating regime of the engine under operating conditions is ordinarily established, and the practicability and limits of thrust regulation by changing the per-second propellant-flow rate to the combustion chamber are ascertained.

Flow-rate characteristics of a ZhRD, in accordance with the pressure of the gases in the combustion chamber, may be constructed in accordance with the data from ground-level tests of the engine on a stand, and analytical calculations for ground-level or arbitrary values of altitude and flight speed.

Flow-rate characteristics for an engine operating on a test stand are usually derived from an invariable weight ratio of the propellant components. Fulfilling this condition, with the changing operating regimes of an engine is, in practice, extremely difficult. Flow-rate characteristics may also be derived for determining the specific thrust of an engine at different weight ratios of the propellant components.

In practice, one may calculate flow-rate characteristics of an engine only approximately, since one cannot precisely evaluate the decrease in the heat-liberation coefficient φ_k of the fuel and the change in the polytropic exponent $\underline{n}$ of the expansion of the gases in the nozzle, because of the change in the per-second flow rate G_s of the propellant to the combustion chamber in relationship to its rated nominal value $G_{s\ r}$. In calculating flow-rate characteristics, the values of φ_k and $\underline{n}$ are assumed as being constant, and equal to their values when the engine is operating in the optimum regime.

Ignoring changes of φ_k and $\underline{n}$, these ratios are found in accordance with the flow-rate characteristics of an engine:

$$\frac{G_s}{G_{s\ r}} = \frac{p_k}{p_{k.r}},$$

whence $C_s = G_{s\ r}(p_k/p_{k.r})$ kg/sec;

$$\frac{p_v}{p_{v.r}} = \frac{p_k}{p_{k.r}},$$

whence $p_v = p_k p_{v.r}/p_{k.r}$ atm abs;

$$\frac{\Delta p_{s.p}}{\Delta p_{s.p.r}} = \left(\frac{G_s}{G_{s\ r}}\right)^2,$$

whence $\Delta p_{s.p} = \Delta p_{s.p.r}(G_s/G_{s\ r})^2 = \Delta p_{s.p.r}(p_k/p_{k.r})^2$, where $\Delta p_{s.p.r}$ and $\Delta p_{s.p}*$ are the summary pressure differences of the fuel in the pressurized feed system of the engine when the chamber is operating in rated and nonrated operating regimes, respectively, in kg/cm^2.

Considering these ratios, the general thrust equation of an engine chamber may be given the form

$$P_s = \frac{w_s}{g} G_s + F_s p_s - F_s p_a = \frac{w_s}{g} G_{s\ r}\frac{p_k}{p_{k.r}} + F_s p_c\frac{p_{v.r}}{p_{k.r}} - F_s p_a =$$

$$= \left(\frac{w_s}{g}\cdot\frac{G_{s\ r}}{p_{k.r}} + F_s\frac{p_{v.r}}{p_{k.r}}\right)p_k - F_s p_a = \left(\frac{P_{pm}}{p_{k.r}} + F_s\frac{p_{v.r}}{p_{k.r}}\right)p_k - F_s p_a.$$

_____

*$[\Delta p_{c.n.p} = \Delta p_{s.p.r} = \Delta p_{summarnyy.\ perepad.\ raschetnyy} =$
$= \Delta p_{summary.\ difference.\ rated}$;
$\Delta p_{c.n} = \Delta p_{summary.\ difference}.$]

Replacing the constant value for a given engine $P_{din}/p_{k.\,r}$ + $F_v p_{v.\,r}/p_{k.\,r}$ = A in this expression, we will finally obtain an equation for computing the thrust of an engine under chosen values of pressure in the combustion chamber:

$$P_s = A p_s - F_s p_s.$$

Since, according to the ratio $p_k = (p_{k.\,r}/G_{s\,r})G_s$ given above, the latter equation may be given the form

$$P_s = A p_s - F_s p_s = A\frac{p_{s\theta}}{G_{s\theta}} G_s - F_s p_s = A'G_s - F_s p_s.$$

where $A' = (P_{din} + F_v p_{v.\,r})/G_{s\,r}$ is a constant magnitude for the given engine.

The per-second propellant-flow rate $G_{s\,r}$ to the combustion chamber when the engine is operating in a rated regime is determined according to the formulas in Section 8 of this chapter.

The thrust of the engine chamber when operating in space, i.e., when $F_v p_a = 0$, is determined in accordance with the formulas:

$$P_p = A p_k \quad or \quad P_p = A'G_s.$$

These formulas show that when the engine is operating in space the dependence of P_p on p_k or G_s is a straight line passing through the beginning of the coordinates, and when operating in an atmosphere it is a straight line arranged below the previous line due to the effect of $F_v p_a$ (Figs. 4.15 and 4.16). One must bear in mind that the flow-rate characteristics of an engine, in real conditions, are an almost straight line only until a propellant-flow rate to the chamber equal to about 30-40% of nominal is achieved. At lower propellant-flow rates, the flow-rate characteristic runs along a sharply increasing curve with its left branch directed downward.

The specific thrust of an engine chamber, with a corresponding value of p_k, is determined in accordance with the formula

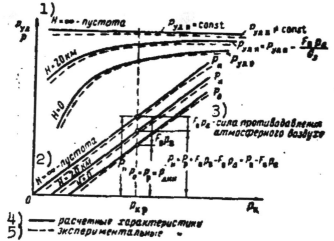

Fig. 4.15. Typical flow-rate characteristics of an engine when operating
in the atmosphere and in space. 1)
H = ∞ — space; 2) H = ∞ — space; 3)
force of the back pressure of the atmospheric air; 4) rated characteristics; 5) experimental characteristics.

$$P_{y\text{д.в}} = \frac{P_{\text{в}}}{G_{\text{в}}} = \frac{Ap_{\text{к}} - F_{\text{в}}p_{\text{в}}}{p_{\text{к}}\dfrac{G_{\text{в}\,\text{р}}}{p_{\text{к.р}}}} = \frac{Ap_{\text{к.р}}}{G_{\text{в}\,\text{р}}} - \frac{F_{\text{в}}p_{\text{к.р}}}{G_{\text{в}\,\text{р}}}\frac{p_{\text{в}}}{p_{\text{к}}}.$$

For a given engine, the relationship $Ap_{k.\,r}/G_{s\ r}$ is of constant
magnitude. Having designated it as $\underline{B}$, and considering that in space
$p_a = 0$, we find the expression for the operation of the engine in
space:

$$P_{y\text{д.в}} = \frac{Ap_{\text{к.р}}}{G_{\text{в}\,\text{р}}} = B = \text{const.}$$

This expression shows that, assuming constant φ_k and $\underline{n}$, according
to the throttle characteristics, a change in specific thrust, depending on p_k or G_s, when the engine is operating in space, is a horizontal line (see Fig. 4.15).

As a consequence of the deterioration in the operating economy of
the engine because of the change in G_s in relationship to $G_{s\ r}$ and the
presence of hydrodynamic losses in the energy of the gases, in view of
the operation of the nozzle under optimum conditions, the real depend-

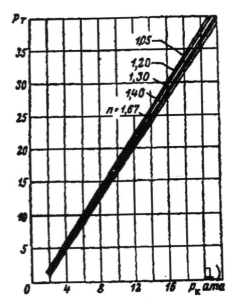

Fig. 4.16. Flow-rate
characteristics of the
A-4 engine when operating
at sea level, computed
for various polytropic
exponents $\underline{n}$. 1) p_k,
atm abs.

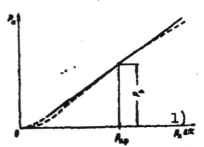

Fig. 4.17. Approximate
curves of rated (solid
line) and experimental
(dotted line) flow-rate
characteristics of an en-
gine when operating at
ground level. 1) p_k,
atm abs.

ence of thrust on pressure in the combustion chamber is a slightly curved
line, almost straight (Fig. 4.17). Since
the curvature of this line is extremely
small, when there is a lack of test-
stand data for an engine, one may fully
rely upon the calculated flow-rate characteristics.

Calculation of a change in an engine's thrust, depending on p_k, may be
carried out by means of the nomogram in
Fig. 4.10 by the following method:

1) for a given value of $f_v =$
$= F_v/F_{kr}$ and a chosen magnitude of $\underline{n}$,
determine, in accordance with the nomo-
gram, the relationship of p_k/p_v and the
thrust coefficient K_p in space;

2) for every chosen value of p_k,
determine $K_n = K_p - f_v p_a/p_k$, and compute
the thrust of the engine in accordance
with the formula $P_n = K_n p_k F_{kr}$.

On the basis of calculating the dependence of P_n and $P_{ud.n}$ on p_k, the already well-known general formula of the
thrust of an engine may be laid down:

$$P_a = p_k F_{kp} \sqrt{\frac{2n^2}{n-1}\left(\frac{2}{n+1}\right)^{\frac{n+1}{n-1}}\left[1-\left(\frac{p_a}{p_k}\right)^{\frac{n-1}{n}}\right]} + F_a(p_a - p_0).$$

For computing the thrust of an engine in accordance with this
formula at chosen values of p_k (for example, with $p_k =$ 5, 10, 15, 20

atm abs, etc.), one must set down the assumed mean values for the polytropic exponent $\underline{n}$ and, by means of the graphs in Fig. 3.9, determine, according to the $f_v = F_v/F_{kr}$ of the given engine, and the chosen value of $\underline{n}$, the relationship of p_k/p_v and then of p_v.

The results of such calculations of the throttle characteristics of an A-4 engine, when operating at sea level with various values of $\underline{n}$, are shown in Fig. 4.16. For this engine, $f_v = 3.42$.

The curves of this figure show that $\underline{n}$ has an insignificant influence on the throttle characteristics of an engine. For example, when $\underline{n}$ is changed from 1 to 1.67, the thrust of an A-4 engine, at a given p_k, is changed by approximately 10%. Changing $\underline{n}$ at an invariable value of p_k causes a decrease in the engine's thrust.

Since, with corresponding values of p_k, it is impossible to evaluate the magnitude of the polytropic exponent $\underline{n}$ of the expansion of the gases in the nozzle precisely, in this case the results of the flow-rate characteristics of the engine are obtained in approximation.

For existing engines, the mean value of $\underline{n}$ may be determined with sufficient accuracy by the magnitudes of p_k and p_v measured during experiments.

For construction of the flow-rate characteristics of an engine chamber, one may use the formulas:

$$P_a = p_k F_{kp} \varphi_c K_{a.r} - F_a p_h; \quad G_s = \frac{G_{sp}}{p_{kp}} p_k \text{ and } P_{yA.s} = \frac{P_k}{G_s}.$$

For calculating the flow-rate characteristics of an engine in accordance with the feed pressure p_p of the propellant components to the combustion chamber, one may use the formula

$$P_a = p_k + \Delta p_{c.s} = p_k + \Delta p_{c.s.p}\left(\frac{p_k}{p_{kp}}\right)^2 = p_k + \Delta p_{c.s.p}\left(\frac{G_s}{G_{sp}}\right)^2.$$

Engines of the final stages of multistage missiles must begin and end their operation in space. This circumstance requires knowledge of

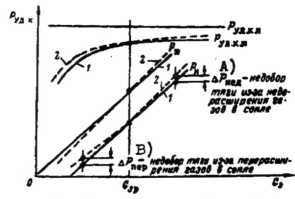

Fig. 4.18. Flow-rate characteristics of an engine with nonregulated and with ideally regulated nozzle in accordance with
altitude. 1) Engine with nozzle
which is not regulated in accordance with flight altitude; 2) engine with ideally regulated nozzle. A) ΔP_{ned} — shortage of
thrust because of underexpansion
of the gases in the nozzle; B)
ΔP_{per} — shortage of thrust because of overexpansion of the
gases in the nozzle.

the flow-rate characteristics of high-altitude and super-high-altitude
engines and the capability of constructing them on the basis of data
from ground-level static tests.

Figure 4.18 shows the flow-rate characteristics of a ZhRD with a
nonregulated high-altitude nozzle in accordance with flow-rate G_s
(solid lines) and with an ideally regulated nozzle in accordance with
the rated conditions of operation (when $p_v = p_a$).

In a given case, when G_s deviates from the rated magnitude $G_{s\ r}$,
nozzle regulation is accomplished by shortening or elongating the
nozzle. Thus, an unregulated engine nozzle operates under nonoptimum
conditions, i.e., the gases overexpand or underexpand in relationship
to the surrounding medium's pressure p_a.

Figure 4.19 shows the flow-rate characteristics of a low-altitude

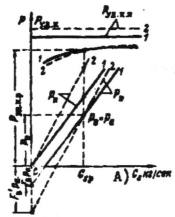

Fig. 4.19. Flow-rate
characteristics of a
low-altitude and of a
high-altitude engine.
1) Low-altitude en-
gine; 2) high-alti-
tude engine. A) G_s,
kg/sec.

and of a high-altitude ZhRD when operating in
the atmosphere and in space (the dashed lines
refer to the high-altitude engine).

Comparing the rated and experimental
flow-rate characteristics of ZhRD is possible
only if they are reduced to the same conditions (preferably to normal conditions).

The following approximate characteristics are also of interest in selecting the
most advantageous engine operating regime
(Fig. 4.20):

1) the dependence of specific fuel consumption C_{ud} and the engine's internal effi-

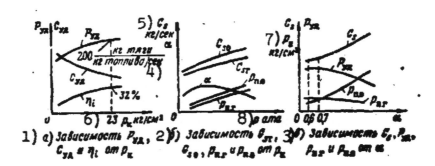

1) a) Зависимость $P_{уд}$, 2) б) Зависимость $G_{ст}$, 3) в) Зависимость G_s, $P_{ш}$,
$C_{уд}$ и η_i от $p_к$ G_{so}, $P_{вr}$ и $P_{вo}$ от $p_к$ $P_{вr}$ и $P_{вo}$ от α

Fig. 4.20. Some engine cycle parameters as
functions of combustion chamber pressure and
the fuel's excess oxidation coefficient. 1)
a) P_{ud}, C_{ud}, and η_1 as functions of p_k; 2)
b) $G_{s\,g}$, $G_{s\,o}$, $P_{p.g}$, and $p_{p.o}$ as functions of
p_k; 3) c) G_s, P_{ud}, $p_{p.g}$, and $p_{p.o}$ as functions
of α; 4) kg-thrust/(kg-fuel/sec); 5) G_s, kg/sec;
6) p_k, kg/cm^2; 7) p_p, kg/cm^2; 8) p, atm abs.

ciency η_1 on combustion chamber pressure p_k;

2) the dependence of the per-second flow rates of the combustible
($G_{s\,g}$) and oxidizer ($G_{s\,o}$) to the combustion chamber, their feed pressures
($p_{p.g}$ and $p_{p.o}$), and the fuel's excess oxidation coefficient α on com-

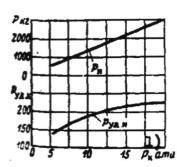

Fig. 4.21. Rated flow-rate engine characteristics (example No. 3). 1) p_k, atm abs.

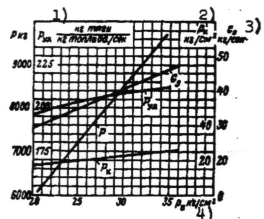

Fig. 4.22. Rated flow-rate engine characteristics. 1) P_{ud}, kg-thrust/(kg-fuel/sec); 2) p_k, kg/cm^2; 3) G_s, kg/sec; 4) p_p, kg/cm^2.

TABLE 4.2

1) p_k, кг/см^2	20	15	10	5
2) P_n кг	3000	2150	1296	445
3) $P_{ud.n}$ $\frac{кг\ тяги}{кг\ топлива/сек}$	221	210	191	131

1) p_k, kg/cm^2; 2) P_n, kg; 3) $P_{ud.n}$, kg-thrust/(kg-fuel/sec).

bustion chamber pressure;

3) the dependence of the per-second fuel-flow rate G_s to the combustion chamber, fuel-component supply pressures $p_{p.g}$ and $p_{p.o}$, and absolute P and specific p_{ud} [sic] engine thrusts on the excess oxidation coefficient α.

The aggregate of the different flow-rate and altitude characteristics permits sufficiently accurate evaluation of ZhRD of any type and design from the viewpoint of per-second fuel-flow rate, thrust, operating efficiency, and those requirements which are demanded for the chamber, fuel-feed system units, and engine operation regulation.

Example 3. Calculate the flow-rate characteristics of an engine when p_k = 20, 15, 10, and 5 kg/cm^2 for a V = 0 and H = 0, if P_r = = 3000 kg, F_v = 407.1 cm^2, $p_{k.r}$ = 20 kg/cm^2, $G_{s\ r}$ = 13.54 kg/sec, and $p_{v.r}$ = 1.00 kg/cm^2.

Solution.

1. At the given initial data for the engine, the flow-rate characteristic computation formulas take the form

$$P_n = 170.3p_k - 407.1 \text{ kg;}$$
$$P_{ud.n} = 251.1 - 602/p_k \text{ kg-thrust/(kg-fuel/sec).}$$

2. The results of the computations for absolute and specific thrust for the given gas pressures in the combustion chamber are reduced in Table 4.2 and presented graphically in Fig. 4.21.

Figure 4.22 shows the flow-rate characteristics of an engine as a function of fuel-feed pressure, with a rated thrust of 8500 kg at ground level and a specific thrust of 213 kg-thrust/(kg-fuel/sec), combustion chamber pressure of 22 atm abs, nozzle outlet-section pressure 1 atm abs, per-second fuel-flow rate to the combustion chamber 39.9 kg/sec, and pressure difference of 7.7 atm abs in the fuel line. The chamber-nozzle outlet-section area is 1052 cm^2.

SECTION 11. REAL ZhRD FLOW-RATE CHARACTERISTICS

Real flow-rate characteristics may be constructed in accordance with the data from ZhRD static firing tests, i.e., considering the factual values of the different additional engine parameters which affect its operation at any instant of time.

During an engine's operation under real conditions, the following parameters, which affect the magnitudes of absolute and specific thrust, are changed:

1) the engine operating components' (fuel and vapor-gas genera-

tion devices) specific weights, caused by their temperature change;

2) the fuel components' pressure head to the pumps in accordance with the degree of their consumption;

3) the ratio of the fuel components (their weight concentration);

4) the gas-vapor pressure to the TNA or the tank pressure in other propellant-feed systems, and similar.

When these parameters are changed, the TNA rpm change accordingly, as well as the working components' per-second flow rate in the engine, combustion chamber pressure, and, consequently, engine thrust.

ZhRD flow-rate and altitude characteristics are usually computed at these parameters' fixed nominal values. They are always differentiated from experimental characteristics. To construct precise engine characteristics by calculations, one must consider an engine's change due to the factors mentioned above; thus, the following system of equations is obtained:

$$
\left.
\begin{aligned}
\Delta G_{s\,o} &= a\Delta p_o + b\Delta p_r + c\Delta \gamma_o + e\Delta \gamma_r + f\Delta G_{nr} + i\Delta c_1; \\
\Delta G_{s\,r} &= a'\Delta p_o + b'\Delta p_r + c'\Delta \gamma_o + e'\Delta \gamma_r + f'\Delta G_{nr} + i'\Delta c_1; \\
\Delta n &= l\Delta G_{s\,o} + m\Delta G_{s\,r} + o\Delta \gamma_o + q\Delta \gamma_r + r\Delta G_{nr} + s\Delta c_1,
\end{aligned}
\right\}
\tag{4.28}
$$

where $\Delta G_{s\,o}$ and $\Delta G_{s\,g}$ are the changes in the per-second flow rate of the oxidizer and the combustible, respectively, in kg/sec; Δp_o and Δp_g are the changes of fuel components' pressure heads to the pumps; $\Delta \gamma_o$ and $\Delta \gamma_g$ are the changes in specific weight of the fuel components; ΔG_{pg}* is the change in gas-vapor flow rate in kg/sec; Δc_1 is the change in nozzle velocity at the outlet from the TNA turbine nozzles, determined by the change in temperature and composition of the working fluid during gas-vapor generation; Δn is the change in TNA rpm; and a,b,c...,a',b',c'... are constant coefficients.

Thus, a change in the per-second fuel-flow rate to the engine

——————
*[$\Delta G_{nr} = \Delta G_{pg} = \Delta G_{parogaz} = \Delta G_{gas\ vapor}$.]

combustion chamber is expressed by the formula

$$\Delta G_s = \Delta G_{s,o} + \Delta G_{s,r} = (a+a')\,\Delta p_0 + (b+b')\,\Delta p_r + (c+c')\,\Delta \gamma_o +$$
$$+ (e+e')\,\Delta \gamma_r + (f+f')\,\Delta G_{nr} + (i+i')\,\Delta c_1. \qquad (4.29)$$

The change in the weight coefficient of the fuel composition will be

$$\Delta \chi = \frac{G_{s,o} + \Delta G_{s,o}}{G_{s,r} + \Delta G_{s,r}} - \chi. \qquad (4.30)$$

where χ is the fuel-composition coefficient when the engine is operating in a nominal regime (when all changes in the additional parameters are equal to zero).

Thus, a change in engine thrust is found by means of the flow-rate characteristics, using the values of ΔG_s and Δ_χ which have been found.

During an engine's flight at various altitudes, with great velocities or accelerations, the number of additional factors affecting the engine's operation increases.

A weapon's acceleration during flight depends basically on the device's weight decrease as a consequence of the consumption of the fuel in the tanks and the decrease in atmospheric air resistance with increased altitude.

SECTION 12. ENGINE ALTITUDE CHARACTERISTICS

For practicable use of an engine, it is important to know how its characteristics are changed by changes in the external operating conditions, i.e., altitude H and flight speed V.

The dependence of absolute P and specific P_{ud} engine thrusts on flight altitude H, at a fixed combustion-chamber pressure p_k and constant flight speed V, is called the altitude characteristics of a ZhRD.

Since an engine may operate in maximum, nominal, and minimal thrust regimes, an engine's altitude characteristics must be con-

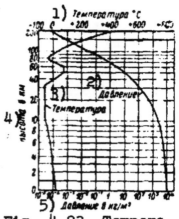

Fig. 4.23. Temperature and pressure in the atmosphere at altitudes of zero to 200 km, 1) Temperature, $^{\circ}$C; 2) pressure; 3) temperature; 4) altitude, km; 5) pressure, kg/m^2.

structed for several combustion-chamber pressures which control these operating regimes.

Constructing the altitude and velocity characteristics of a ZhRD by the experimental method is attended with extremely great difficulties, since the use of very complicated equipment is required for this — special wind tunnels and pressure chambers, and providing actual operational tests for the engine installed in a missile provided with apparatus capable of measuring the required parameters with sufficient accuracy. Therefore these ZhRD characteristics are usually constructed by the calculation method. Thus, precise calculation of these characteristics is made extremely complicated by the impossibility of precisely evaluating the change in atmospheric air pressure around the engine and behind its nozzle during changes in altitude and flight speed.

The pressure p_a of the surrounding medium, into which gases flow from the engine nozzle, is the basic external factor affecting the magnitudes of the engine's absolute and specific thrusts. This pressure is changed as a function of the change in altitude and flight speed.

For constructing the altitude characteristics of a ZhRD, ignoring the change in air pressure behind the engine nozzle as a consequence of the change in flight speed (the engine's ideal altitude characteristics), we may take the values of p_a, as a function of flight altitude, from Table 4.3, the Standard International Atmosphere, which has been computed up to an altitude of 75 km.

TABLE 4.3

Standard International Atmosphere

Высота км 1)	Давление кг/см² 2)	Темпера- тура T °K 3)	Высота км 1)	Давление кг/см² 2)	Темпера- тура T °K 3)
0	1,033	288	16	0,113	216,5
1	0,917	281,5	17	0,089	216,5
2	0,811	275	18	0,076	216,5
3	0,715	268,5	19	0,065	216,5
4	0,620	262	20	0,056	216,5
5	0,550	255,5	21	0,048	216,5
6	0,481	249	22	0,041	216,5
7	0,418	242,5	23	0,035	216,5
8	0,363	236	24	0,030	216,5
9	0,314	229,5	25	0,025	216,5
10	0,269	223	26	0,020	216,5
11	0,231	216,5	30	0,0124	
12	0,197	216,5	40	0,003	
13	0,168	216,5	50	0,00093	
14	0,143	216,5	60	0,00031	
15	0,122	216,5	75	0,000031	

1) Altitude, km; 2) pressure, kg/cm^2; 3) temperature T, °K.

At altitudes above 25 km, the value of p_a may be evaluated according to the curves in Fig. 4.23. These curves show that at great altitudes p_a is so small that it may be neglected; if we consider that $p_a = 0$ at an altitude of 20 km, the error in computing thrust would not exceed 1%. Therefore, engine altitude characteristics may be computed with sufficient precision from the data in Table 4.3 alone.

Relative ZhRD thrust increase as altitude increases depends on the designed altitude of the nozzle, and in contemporary engines, within the limits of atmospheric pressure change from one atmosphere to space, it may attain 10-20%.

The determination of p_a as a function of flight speed presents great difficulties and is possible only on the basis of the weapon's aerodynamic design, allowing for the weapon's shape.

Figure 4.24 shows the altitude characteristics for an A-4 engine,

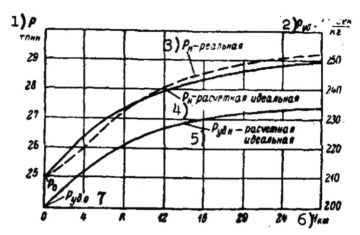

Fig. 4.24. A-4 engine altitude characteristics (dashed line shows real
characteristics). 1) P, tons; 2) $P_{ud.n}$,
kg-sec/kg; 3) P_n — real; 4) P_n — rated
ideal; 5) $P_{ud.n}$ — rated ideal; 6) H,
km; 7) $P_{ud\ 0}$.

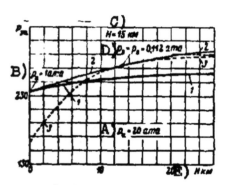

Fig. 4.25. Altitude char-
acteristics of low-alti-
tude and high-altitude
engines, and of an engine
with ideal nozzle, regu-
lated in accordance with
altitude. 1) Low-altitude
engine; 2) engine with
ideal nozzle, regulated
in accordance with alti-
tude; 3) high-altitude
engine. A) p_k = 20 atm
abs; B) p_v = 1 atm abs;
C) H = 15 km; D) p_v = p_a =
= 0.112 atm abs; E) H, km.

where F_v = 4295 cm^2, p_v = 0.85 atm abs,
P_0 = 25 tons, and P_{ud0} = 200 kg-thrust
per kg-fuel/sec.

The curves in this graph show that
at an altitude of 28 km the engine develops approximately 15.5% greater
thrust than at ground level, and, when operating in space, 17% greater thrust. At
a fixed flow rate to the combustion
chamber, not only these percentages grow,
but the engine's specific thrust as well.

Figure 4.25 shows the approximate
rated altitude characteristics for low-
altitude and high-altitude engines, also
that of an engine with an ideal nozzle,
regulated in accordance with flight altitude, as constructed in accordance

162

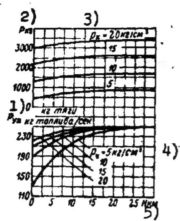

Fig. 4.26. Rated engine altitude characteristics (see example No. 4). 1) P_{ud}, kg-thrust/kg-fuel per sec; 2) P, kg; 3) p_k = 20 kg/cm²; 4) p_k = 5 kg/cm²; 5) H, km.

with the values of p_a taken from Table 4.7 [sic]. All these engines are identical, i.e., they all operate on the same fuel and at the same combustion-chamber pressure.

The curves of this graph show that:

1) a high-altitude engine has a relatively small thrust at ground level, which gives the weapon poor launching properties;

2) as flight altitude increases, the absolute thrust of a high-altitude engine increases more intensively than that of a low-altitude engine, in direct proportion to the engine's design altitude;

3) an engine with an ideal nozzle, regulated in accordance with flight altitude, has, relatively speaking, more favorable characteristics, which shows the necessity of making such an engine with a regulated design altitude, even if it has only one or two fixed positions.

Besides high operating economy, an engine must provide the weapon with a good takeoff, i.e., its thrust at ground level must exceed the missile's blastoff weight by a given number. An engine's relative thrust coefficient b = P_0/G_0 is a function of the tactical designation of the missile. Calculations show that as combustion-chamber pressure rises, the increase in an engine's absolute and specific thrusts becomes less and less significant as flight altitude increases. One should therefore bear in mind that specific thrust will be changed as a function of the engine's absolute thrust, since G_s is independent of p_a and remains constant in calculating the external characteristics.

<u>Example 4.</u> Calculate the altitude characteristics for an engine

TABLE 4.4

1) H, км	0	5	10	20	30
2) p_a, кг/см²	1,033	0,550	0,269	0,055	0,011

1) H, km; 2) p_a, kg/cm².

TABLE 4.5
Values of $P_n/P_{ud.n}$ in Accordance with Flight Altitude

Давление в камере сгорания $P_к$ кг/см² 1)	2) Высота полета H в км				
	0	5	10	20	30
20	$\dfrac{2986}{220}$	$\dfrac{3198}{235}$	$\dfrac{3298}{243}$	$\dfrac{3385}{250}$	$\dfrac{3404}{251}$
15	$\dfrac{2137}{210}$	$\dfrac{2333}{229}$	$\dfrac{2448}{240}$	$\dfrac{2535}{249}$	$\dfrac{2553}{251}$
10	$\dfrac{1283}{189}$	$\dfrac{1479}{218}$	$\dfrac{1594}{235}$	$\dfrac{1681}{248}$	$\dfrac{1699}{251}$
5	$\dfrac{432}{127}$	$\dfrac{628}{185}$	$\dfrac{743}{210}$	$\dfrac{830}{245}$	$\dfrac{848}{250}$

1) Pressure p_k in combustion chamber, kg/cm²; 2) flight altitude H, km.

when p_k = 20, 15, 10, and 5 kg/cm² for flight altitudes H = 0, 5, 10, 20, and 30 km, if P_r = 3000 kg, F_v = 407.1 cm², p_k = 20 kg/cm², $G_{s\ r}$ = 13.54 kg/sec, and $p_{v.r}$ = 1.00 kg/cm².

Solution.

1. We determine the absolute and specific thrusts of the engine in accordance with the formulas:

$$P_n = AP_k - F_v p_a \text{ and } P = \frac{P_a}{G_s} = B - C\frac{p_a}{p_k},$$

which, at the given parameters, is reduced to the form:

a) when p_k = 20 kg/cm² P_n = 3407-407.1 p_a;

$$P_{ud.n} = 251.1-30.1 \; p_a;$$

b) when $p_k = 15$ kg/cm^2 $P_n = 2557-407.1 \; p_a;$

$$P_{ud.n} = 251.1-20.1 \; p_a;$$

c) when $p_k = 10$ kg/cm^2 $P_n = 1703-407.1 \; p_a;$

$$P_{ud.n} = 251.1-60.2 \; p_a;$$

d) when $p_k = 5$ kg/cm^2 $P_n = 852-407.1 \; p_a;$

$$P_{ud.n} = 251.1-120.4 \; p_a,$$

where p_a is in kg/cm^2 and $P_{ud.n}$ is in kg-thrust/(kg-fuel/sec).

2. The atmospheric pressure for the given altitudes, taken from Table 4.7 [sic], is reduced in Table 4.4.

3. The results of the computations of absolute and specific thrusts, at the given flight altitudes, are shown in Table 4.5 and presented graphically in Fig. 4.26.

SECTION 13. REAL ZhRD ALTITUDE CHARACTERISTICS

In the previous section, we considered the change in an engine's absolute and specific thrusts as a function of the natural pressure change in the unperturbed atmospheric air as a function of flight altitude. Under real engine operating conditions, the pressure of the surrounding medium behind the chamber nozzle is always less than the pressure of the unperturbed air. This phenomenon is caused by the rarefication developed behind the engine nozzle as the missile flies with great velocities and accelerations, and affects the magnitude of the engine's thrust. Therefore, in precise calculation of an engine's altitude characteristics, one must consider its specific external operating conditions.

In real operating conditions, an engine's absolute thrust depends on the following factors:

1) the weapon's flight altitude, speed, and acceleration;

2) the fuel tanks' pressurization system;

3) the engine's propellant-feed system;

4) the specific weight and level of the fuel components in the tank, etc.

Some of these factors tend to increase, and some to decrease, the engine's absolute thrust. For example, as the flight altitude and speed of a weapon of the A–4 type are increased, the pressure of the air around the engine decreases in relationship to the atmospheric pressure, and in the presence of a great flight speed (above Mach number 2.6) the rarefication formed behind the engine nozzle may even attain absolute vacuum, as a result of which the following things take place.

1. The expansion ratio of the gases behind the engine nozzle increases as a consequence of the decreased back pressure p_a, because of which the engine thrust increases by means of the increase in the statistic member $P_{st} = F_v (p_v - p_a)$ (subscript st – statistic).

2. The expansion ratio of the turbopump unit exhaust gases increases as they flow out into the surrounding medium, and volume of turbopump unit exhaust gases increases as a consequence of the decreased back pressure p_a (if a TNA is part of the engine propellant-feed system and the turbine exhaust gases continue to expand as they flow out into the surrounding medium and develop a thrust).

3. The on-board gas pressure on the engine fuel-system pressure reducer is reduced at the outlet as a consequence of the decreased pressure p_a of the surrounding medium on the reducer diaphragm, because of which the gas-vapor feed to the turbine is reduced, turbine rpm drops, and the pumps' capacity is lowered, which leads to a decrease in engine thrust. An analogous effect on engine thrust is possible when other instruments which are sensitive to pressure changes

in the surrounding medium are installed.

4. The pressure for the engine fuel-pump suction is changed as a consequence of a change in fuel-tank pressurization (if pressurization by velocity head or from an on-board source of compressed gas is used), because of which changes occur in the per-second fuel-flow rate to the combustion chamber and, consequently, in the engine thrust.

5. The fuel components' pressure head to the pumps is increased by means of the inertial forces developed as the weapon is accelerated, in direct proportion to the specific weight and the level of the fuel components in the tanks before they reach the pumps, which, similar to item No. 4, changes the engine thrust.

6. The fuel components' weight ratio as they are fed to the engine combustion chamber is decreased under the influence of the factors mentioned in items 3-5, because of which the fuel's thermal efficiency is changed and, consequently, the engine thrust as well.

To calculate the effect of all the factors enumerated above on thrust magnitude and to construct the real altitude characteristics, it is necessary to have the following basic characteristics:

a) engine combustion chamber and nozzle — the pressure of the gases in the chamber and at the outlet from the nozzle, the area of the nozzle outlet section;

b) propellant-feed system — turbines, pumps, pressure regulators, and the like, and

c) weapon — flight altitude, speed and acceleration, and angle of shift in accordance with time, as the same engine may show different real characteristics when installed in different missiles.

In practice, this problem is brought to compilation by simultaneous approximate solution of several equations, considering the effect of the factors enumerated above on the engine's operation.

The most important of these factors is the pressure of the atmospheric air behind the engine nozzle; therefore, its magnitude must be evaluated by computing with the specific engine operating conditions when designing a ZhRD.

An engine's real altitude characteristics, i.e., allowing for the effect of all the factors enumerated above on engine operation, is expressed by the equation

$$P_{z\,H} = P_{\phi}\left(1 + \frac{\Delta G_z}{G_z}\right) + F_z(P_z - p_z) + \Delta P_{\text{THA}} - \Delta P_{py\pi}. \tag{4.31}$$

where P_r is the engine thrust at ground level at the rated nominal regimes; ΔP_{TNA} is the thrust developed by TNA exhaust gases; ΔP_{rul} is the engine thrust loss due to gas rudders; $\Delta G_s = \Delta G_{s\,g} + \Delta G_{s\,o}$ is the change in per-second fuel flow to the combustion chamber in kg/sec as the consequence of:

a) the effect of the pressure change in the surrounding air on the pressure reducer's operation, determined, in turn, by the operating regimes of the PGG and TNA (if such devices are included in the engine propellant-feed system);

b) the effect of variation in velocity-head fuel-tank pressurization on the pumps' operation; as flight altitude increases, velocity-head pressurization decreases, and as flight speed increases, velocity-head pressurization increases;

c) the effect of the change in fuel component level in the tanks as a consequence of their consumption;

d) the effect of the changing fuel-component pressure head in the tanks because of the inertial forces developed during the weapon's acceleration, and other reasons.

Figure 4.24 shows the results of the calculation of an A-4 engine's real altitude characteristics when it is installed in a long-range missile.

It is obvious that the same engine may have different real altitude characteristics, depending on the particulars of the aircraft in which it is installed. A change in construction, fuel-tank volume, takeoff velocity or acceleration, head of fuel-tank pressurization, or the like leads unavoidably to a change in the fuel components' feed pressure to the combustion chamber, and, consequently, change in engine thrust.

The construction of real ZhRD altitude characteristics for aircraft is closely connected with the calculation of the trajectory elements for these craft. Therefore, the engine altitude characteristics and the aircraft flight trajectory must be determined by simultaneous calculations.

The curves in Fig. 4.24 show that the real altitude characteristics of an engine differ very little (by 1-3%) from the ideal characteristics (dashed line) when computed under the same initial conditions in conformance to the simplified formula which considers only the normal pressure of the unperturbed air behind the engine nozzle in accordance with flight altitude. Bearing this circumstance in mind, one may use the simplified thrust formula for calculating altitude characteristics.

SECTION 14. SELECTION OF OPTIMUM ENGINE-CHAMBER NOZZLE DESIGN ALTITUDE

One of the basic parameters which directly affects the magnitude of an engine's absolute thrust is the pressure of the surrounding medium, which may change during an aircraft's flight as a function of flight altitude and speed.

The majority of aircraft engines operate without a change of the gas pressure in the combustion chamber as altitude increases, but since atmospheric pressure p_a is thus decreased, the necessity arises

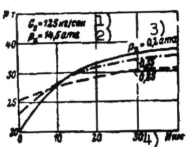

Fig. 4.27. Rated altitude characteristics of an A-4 engine with different conventional nozzle lengths. 1) G_s = 125 kg/sec; 2) p_k = 14.5 atm abs; 3) p_v = 0.2 atm abs; 4) H, km.

for craft of this type to use high-altitude engines, in which, when operating at ground level, the gas pressure p_v in the nozzle outlet section is somewhat less than the pressure p_a of the atmospheric air.

If the magnitude of p_v is properly selected, the engine may develop the greatest mean specific thrust $P_{ud.sr}$ in accordance with its flight trajectory and give the aircraft, other things being equal, the maximum flight range.

The results of the computations and construction of the altitude characteristics for an A-4 engine with nozzles of different design altitudes, under otherwise equal conditions (see Fig. 4.27), show that:

1) an engine of low design altitude (p_v = 0.85 atm abs), when operating at ground level, develops a relatively greater thrust than a high-altitude engine (see the curve for p_v = 0.85 atm abs), but then as it rises in altitude its thrust increases less rapidly, and, as a result, at a high altitude becomes relatively low; because of this $P_{ud.sr}$ will also be low;

2) an engine of excessively high design altitude develops a relatively great thrust at considerable altitudes, but it has an extremely small thrust at ground level and at low flight altitudes (see the curve for p_v = 0.2 atm abs), as a result of which the $P_{ud.sr}$ obtained is also small;

3) an engine of moderate design altitude, obviously, will have a relatively great $P_{ud.sr}$ (see the curve for a p_v = 0.35 atm abs);

4) the A-4 missile's increase in flight range when p_v = 0.35

atm abs amounts to about 8% of the range obtained when $p_v = 0.85$ atm abs.

Since the operation time of a ZhRD in different regimes according to thrust and flight altitude depends basically on the aircraft's flight trajectory, it is necessary to choose a nozzle design altitude for an engine being designed only for the specific reaction device in which the engine will be installed.

In designing an engine it is necessary to select a p_v that, other things being equal, will give the greatest $P_{ud.sr}$ and, consequently, the greatest missile range.

The optimum engine-nozzle design altitude depends on the aircraft's flight trajectory, which shows that there is some difficulty in computing $p_{v.opt}$,* which is required in advance in order to know the flight trajectory of the installation being designed. The greater the aircraft's perfection from the design viewpoint, i.e., the smaller its relative final weight μ_{kon}** (for the A-4 missile we have a $\mu_{kon} = G_k/G_o = 0.32$) and the greater the engine's specific thrust, the less the value of $p_{v.opt}$ will be and the greater the effect that may be expected from using a nozzle of optimum design altitude.

The use of high-altitude nozzles provides a considerable increase in aircraft range in comparison to a low-altitude nozzle, in direct proportion to the aircraft's degree of perfection.

In selecting a nozzle design altitude for an engine being designed, one should bear in mind that:

1) increasing the design altitude of the nozzle increases its dimensions and weight, because of which difficulties arise in cooling its greater surface, and also the expenditures of energy in climbing

---

*[$p_{в.опт} = p_{outlet.optimum}$.]
**[$\mu_{кон} = \mu_{kon} = \mu_{konechnyy} = \mu_{final}$.]

and acceleration increase sharply because of the engine's increase in
weight and the decrease in p_v;

2) the engine thrust at ground level is decreased considerably,
which greatly impairs the missile's launching properties;

3) when the gases in the nozzle overexpand up to a pressure of
less than $p_v \approx 0.3 p_a$ the gas flow becomes detached from the nozzle
walls, because of which compression waves appear, which decrease the
engine's specific thrust considerably.

In view of these reasons, and also because the launching thrust
of fast-climbing, heavy missiles of the A-4 type must, in practice, be
not less than double the weight of the missile at blastoff, some pressure increase in the nozzle outlet section relative to its optimum
value is advisable. Thus, the engine nozzle is shortened, its weight
is decreased, and launching thrust increases. The latter circumstance
indicates the necessity for comparative evaluation of the effect of
the engine's weight and specific thrust on the missile's range (to attain the greatest range, one must determine the weight equivalent of
the thrust).

In designing an engine, we often base the magnitude of the
pressure in the nozzle outlet section on statistical data
(considering the missile's designation).

At the present time, for single-stage missile engines, it is assumed that in the nozzle outlet section the gas pressure $p_v = 0.6-0.85$
atm abs.

Sometimes the magnitude of p_v is assumed as equal to the atmospheric air pressure p_a at the altitude which the missile reaches after
expending half of the supply of fuel in the tanks. One cannot consider
such a solution to this problem as valid.

For engine nozzles intended for the second and following stages

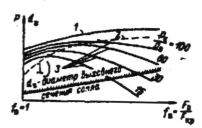

Fig. 4.28. Absolute engine thrust as a function of the dimensionless area f_v of a nozzle cross section. 1) d_v — diameter of nozzle outlet section.

of a multistage missile, in order not to increase the nozzle outlet section by a considerable degree, one may assume p_v = 0.1-0.3 atm abs.

As we select the magnitude of p_v it is necessary to consider the real engine design altitude, the missile's designation, the character of its flight trajectory, nozzle cooling system, the effect of increasing the engine's weight on the missile's range because of the change in p_v, and a number of other factors.

Figure 4.28 shows the curves for the absolute thrust P of an engine chamber as a function of the nozzle's dimensionless area f_v = F_v/F_{kr} at various operating conditions. Curve 1 shows the change in P as a function of the change in f_v when the engine is operating in space; as f_v increases, the magnitude of P increases.

The dashed curve 2 shows the change in P as a function of f_v when the engine is operating continuously at the rated regime; this regime is maintained by changing the per-second fuel-flow rate to the combustion chamber.

The other curves 3 show the change in P as a function of the different ratios of gas expansion within the nozzle; for example ε_s = p_k/p_v = 15, 30, 60, and 100. Each of these curves has its maximum when p_v = p_a, i.e., when the corresponding curve intersects with curve 2.

The curves of this graph show that an engine with a small pressure p_k in the combustion chamber is more sensitive to a change in the nozzle length than when p_k is great. They also indicate that engines with large values of p_k should be made as low-altitude engines, since by this means they will have a relatively small loss of thrust.

SECTION 15. REGULATING AN ENGINE-CHAMBER NOZZLE'S DESIGN ALTITUDE

An engine with a chamber nozzle which was ideally regulated to conform with flight altitude [i.e., a nozzle which was elongated in accordance with flight altitude to maintain the engine's optimum rated operating conditions $(p_v = p_a)$] would have the best altitude characteristics.

This shows the requirement for developing engine chambers with nozzles which are regulated in accordance with flight altitude, even if they have only one or two fixed positions, which, other things being equal, considerably increases a missile's range without decreasing its launching thrust as compared to a missile with a low-altitude engine.

Figure 4.29 shows the altitude characteristics of an engine:

1) with normal ground-level nozzle (curve 1-4-2);

2) with an oversized ground-level nozzle (curves 3-4-5-6-7 and 8-6-9-10);

3) with an ideal nozzle, regulated in accordance with altitude (curve 1-5-9-11);

4) with a two-stage regulated nozzle (curve 1-4-5-6-9-10), with the first stage of regulation switched into operation at point 4 and the second at point 6.

The curves of these altitude characteristics show that the use of nozzle design-altitude regulation permits:

1) increasing the engine's launching thrust;

2) increasing the engine's operating economy, determined by the magnitude of $P_{ud.sr}$, and

3) increasing the aircraft's range (by means of the first two factors).

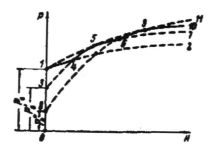 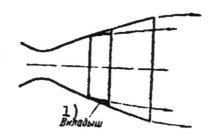

Fig. 4.29. Character of two-stage regulation of engine-nozzle design altitude.

Fig. 4.30. Simplified diagram of single-stage regulation of engine-nozzle design altitude. 1) Liner.

Regulating the nozzle's design altitude as altitude increases is most effective for long-range and super-long-range missiles, whose powered trajectory is close to vertical.

Practical accomplishment of chamber-nozzle design-altitude regulation, even in the simplest form (with one or two fixed values of $f_v = F_v/F_{kr}$), is extremely complicated. The engine nozzle usually operates under high temperature conditions; therefore it is very difficult to regulate it and cool it at the same time.

One-stage nozzle design-altitude regulation may be accomplished without great design difficulties by the use of a special nozzle liner (Fig. 4.30). During the missile's launching and as it climbs to a predetermined flight altitude, this liner will detach the stream of gases from the engine chamber nozzle walls, and therefore it will operate as a normal ground-level nozzle. After a certain altitude is reached, the nozzle liner burns up or is mechanically ejected and the nozzle begins to operate as a high-altitude nozzle.

In practice, other methods of one-stage engine-nozzle design-altitude regulation are possible.

APPENDICES

APPENDIX I

ENERGY CONTENT OF PRODUCTS OF FUEL COMBUSTION I' kcal/kmole*

$T°K$	CO_2	H_2O	CO	OH	NO	H_2
300	−93990,0	−57731,4	−26375,1	10107,9	21645,2	46,9
400	−93049,8	−56919,4	−25676,8	10815,0	22359,2	741,4
500	−92021,8	−56087,9	−24970,6	11520,4	23082,3	1439,9
600	−90922,2	−55231,2	−24251,1	12226,2	23819,6	2139,0
700	−89763,3	−54341,9	−23514,8	12932,3	24575,0	2841,9
800	−88555,2	−53423,8	−22760,8	13643,8	25348,0	3547,8
900	−87304,7	−52478,7	−21990,0	14362,8	26138,9	4258,4
1000	−86022,6	−51505,4	−21204,1	15091,4	26944,4	4976,2
1100	−84709,6	−50505,1	−20404,6	15830,0	27762,9	5702,7
1200	−83371,8	−49477,2	−19593,3	16579,6	28592,1	6438,5
1300	−82013,4	−48423,2	−18771,7	17340,3	29430,5	7184,3
1400	−80637,2	−47344,3	−17941,0	18112,0	30276,7	7940,5
1500	−79245,6	−46242,2	−17102,6	18894,4	31129,6	8707,3
1600	−77840,4	−45118,3	−16257,6	19686,8	31988,3	9484,4
1700	−76423,0	−43974,5	−15406,8	20488,7	32852,0	10271,7
1800	−74995,0	−42812,4	−14550,7	21299,6	33720,1	11068,8
1900	−73557,4	−41633,7	−13690,1	22118,9	34592,1	11875,2
2000	−72111,2	−40440,1	−12825,5	22945,9	35467,6	12690,4
2100	−70657,4	−39232,8	−11957,3	23780,2	36346,2	13514,0
2200	−69196,4	−38013,1	−11085,9	24621,3	37227,6	14345,5
2300	−67729,0	−36782,1	−10211,5	25468,6	38111,6	15184,5
2400	−66255,6	−35540,7	−9334,4	26321,9	38997,9	16030,8
2500	−64776,6	−34289,9	−8454,9	27180,6	39886,4	16883,9
2600	−63292,4	−33030,3	−7573,2	28044,5	40776,8	17743,5
2700	−61803,2	−31762,6	−6689,5	28913,1	41669,1	18609,5
2800	−60309,3	−30487,6	−5803,9	29786,4	42563,2	19481,3
2900	−58811,1	−29205,7	−4916,6	30664,0	43458,9	20358,9
3000	−57308,6	−27917,4	−4027,6	31545,8	44356,3	21242,1

---

* G. B. Sinyarev and M. V. Dobrovol'skiy. Zhidkostnyye raketnyye dvigateli [Liquid-propellant Rocket Engines], Oborongiz [State Publishing House of the Defense Industry], 1957.

O_2	N_2	H	O	N	$T°$ K
47,9	47,6	52115,6	59183,1	85591,6	300
758,0	744,9	52612,4	59701,0	86088,4	400
1489,1	1447,5	53109,2	60211,5	86585,2	500
2244,4	2160,5	53606,0	60717,8	87082,0	600
3022,3	2887,8	54102,8	61221,6	87578,8	700
3819,9	3631,1	54599,6	61723,7	88075,6	800
4634,0	4390,5	55096,4	62224,8	88572,4	900
5461,5	5165,0	55593,2	62724,9	89069,2	1000
6300,3	5953,2	56090,0	63224,5	89566,0	1100
7148,6	6753,7	56586,8	63723,7	90062,8	1200
8005,1	7565,1	57083,6	64222,5	90559,6	1300
8869,0	8386,0	57580,4	64721,0	91056,4	1400
9739,6	9215,1	58077,2	65219,3	91553,2	1500
10616,5	10051,6	58574,0	65717,4	92050,0	1600
11499,4	10894,5	59070,8	66215,4	92546,8	1700
12388,1	11743,0	59567,6	66713,2	93043,6	1800
13282,6	12596,6	60064,4	67211,1	93540,5	1900
14182,7	13454,7	60561,2	67708,8	94037,4	2000
15088,3	14316,8	61058,0	68206,6	94534,3	2100
15999,5	15182,5	61554,8	68704,4	95031,3	2200
16916,1	16051,5	62051,6	69202,3	95528,5	2300
17838,2	16923,4	62548,4	69700,4	96025,8	2400
18765,7	17798,0	63045,2	70198,6	96523,4	2500
19698,4	18675,1	63542,0	70697,1	97021,4	2600
20636,4	19554,5	64038,3	71195,9	97519,8	2700
21579,4	20436,0	64535,6	71695,0	98018,9	2800
22527,3	21319,5	65032,4	72194,6	98518,6	2900
23480,0	22204,7	65529,2	72694,8	99019,2	3000

$T°$ K	CO_2	H_2O	CO	OH	NO	H_2
3100	—55802,6	—26623,4	—3136,0	32431,6	45255,1	22130,6
3200	—54292,7	—25324,1	—2244,0	33321,1	46155,4	23024,4
3300	—52779,2	—24020,0	—1350,7	34214,2	47057,1	23923,3
3400	—51262,4	—22711,4	—456,2	35110,7	47960,2	24827,2
3500	—49742,4	—21398,7	+439,6	36010,6	48864,5	25736,0
3600	—48219,2	—20082,2	1336,4	36913,7	49770,0	26649,6
3700	—46693,2	—18762,1	2234,4	37819,9	50676,7	27568,0
3800	—45164,4	—17438,7	3133,4	38729,1	51584,6	28491,0
3900	—43632,8	—16112,2	4033,4	39641,2	52493,6	29418,5
4000	—42098,6	—14782,8	4934,4	40556,0	53403,8	30350,5
4100	—40561,8	—13450,7	5836,3	41473,6	54315,0	31286,9
4200	—39022,6	—12116,1	6739,2	42393,8	55227,3	32227,6
4300	—37480,9	—10779,1	7643,0	43316,6	56140,6	33172,5
4400	—35936,8	—9439,9	8547,7	44241,9	57054,9	34121,7
4500	—34390,5	—8098,6	9453,2	45169,7	57970,2	35074,9
4600	—32841,8	—6755,2	10359,5	46099,9	58886,5	36032,1
4700	—31291,0	—5409,8	11266,5	47032,5	59803,8	36993,4
4800	—29737,8	—4062,4	12174,3	47967,5	60722,1	37958,6
4900	—28182,6	—2713,2	13082,9	48904,8	61641,4	38927,7
5000	—26625,0	—1362,0	13992,1	49844,3	62561,7	39900,5
5100	—25065,4	—9,0	14902,1	50786,0	63483,0	40877,2
5200	—23503,4	+1346,0	15812,7	51729,7	64405,3	41857,6
5300	—21939,4	2702,8	16724,1	52675,6	65328,5	42841,6
5400	—20373,0	4061,4	17636,1	53623,5	66252,6	43829,3
5500	—18804,6	5422,0	18548,7	54573,4	67177,7	44820,5
5600	—17233,8	6784,3	19462,1	55525,3	68103,8	45815,1
5700	—15661,0	8148,5	20376,1	56479,1	69030,9	46813,5
5800	—14085,8	9514,5	21290,7	57434,9	69958,9	47815,1
5900	—12508,6	10882,3	22206,0	58392,5	70887,8	48820,1
6000	—10929,0	12251,9	23121,9	59351,9	71817,7	49828,5

181

O_2	N_2	H	O	N	$T°$ K
24437,3	23091,6	66026,0	73195,5	99520,9	3100
25399,1	23980,2	66522,8	73696,8	100023,3	3200
26365,2	24870,2	67019,6	74198,9	100528,0	3300
27335,5	25761,6	67516,4	74701,8	101033,9	3400
28309,7	26654,3	68013,2	75205,4	101541,6	3500
29287,8	27548,3	68510,0	75710,0	102051,4	3600
30269,5	28443,5	69006,8	76215,5	102563,3	3700
31254,7	29339,3	69503,6	76722,0	103077,7	3800
32243,2	30237,2	70000,4	77229,5	103594,8	3900
33234,8	31135,6	70497,2	77738,1	104114,7	4000
34229,4	32035,0	70994,0	78247,8	104637,7	4100
35226,8	32935,4	71490,8	78758,6	105164,0	4200
36226,8	33836,6	71987,6	79270,6	105693,8	4300
37229,3	34738,8	72484,4	79783,8	106227,3	4400
38234,1	35641,9	72981,2	80298,1	106764,6	4500
39241,2	36545,8	73478,0	80613,7	107305,9	4600
40250,4	37450,6	73974,8	81330,5	107851,4	4700
41261,6	38356,2	74471,6	81848,5	108401,1	4800
42274,6	39262,6	74968,4	82367,7	108955,4	4900
43289,3	40169,8	75465,2	82888,1	109514,1	5000
44305,7	41077,8	75962,0	83409,7	110077,6	5100
45323,7	41986,5	76458,8	83932,5	110645,8	5200
46343,1	42896,1	76955,6	84456,6	111218,9	5300
47363,9	43806,3	77452,4	84981,8	111796,9	5400
48386,0	44717,4	77949,2	85508,2	112380,0	5500
49409,9	45629,1	78446,0	86035,8	112968,1	5600
50433,8	46541,6	78942,8	86564,6	113561,4	5700
51459,4	47454,9	79439,6	87094,5	114159,9	5800
52485,9	48368,8	79936,4	87625,5	114763,7	5900
53513,4	49283,5	80433,2	88157,7	115372,8	6000

APPENDIX II

CONSTANTS OF CHEMICAL EQUILIBRIUM FOR PRODUCTS OF FUEL COMBUSTION

T° K	$K_{p1} = \dfrac{p_{CO} p_{O_2}^{1/2}}{p_{CO_2}}$	$K_{p2} = \dfrac{p_{H_2} p_{O_2}^{1/2}}{p_{H_2O}}$	$K_{p4} = \dfrac{p_{CO} p_{H_2O}}{p_{CO_2} p_{H_2}}$	$K_{p3} = \dfrac{p_{OH} p_{H_2}^{1/2}}{p_{H_2O}}$
300	$0,1825 \cdot 10^{-44}$	$0,1637 \cdot 10^{-39}$	$0,1115 \cdot 10^{-4}$	$0,5140 \cdot 10^{-46}$
400	$0,3895 \cdot 10^{-32}$	$0,5759 \cdot 10^{-29}$	$0,6764 \cdot 10^{-3}$	$0,1237 \cdot 10^{-32}$
500	$0,9886 \cdot 10^{-25}$	$0,1302 \cdot 10^{-22}$	$0,7593 \cdot 10^{-2}$	$0,3518 \cdot 10^{-26}$
600	$0,8624 \cdot 10^{-20}$	$0,2333 \cdot 10^{-18}$	$0,3696 \cdot 10^{-1}$	$0,3400 \cdot 10^{-21}$
700	$0,2900 \cdot 10^{-16}$	$0,2614 \cdot 10^{-15}$	$0,1109$	$0,1265 \cdot 10^{-17}$
800	$0,1277 \cdot 10^{-13}$	$0,5156 \cdot 10^{-13}$	$0,2475$	$0,6119 \cdot 10^{-15}$
900	$0,1445 \cdot 10^{-11}$	$0,3185 \cdot 10^{-11}$	$0,4537$	$0,7568 \cdot 10^{-13}$
1000	$0,6331 \cdot 10^{-10}$	$0,8728 \cdot 10^{-10}$	$0,7254$	$0,3604 \cdot 10^{-11}$
1100	$0,1389 \cdot 10^{-8}$	$0,1314 \cdot 10^{-8}$	$1,0560$	$0,8519 \cdot 10^{-10}$
1200	$0,1814 \cdot 10^{-7}$	$0,1267 \cdot 10^{-7}$	$1,4320$	$0,1193 \cdot 10^{-8}$
1300	$0,1591 \cdot 10^{-6}$	$0,8648 \cdot 10^{-7}$	$1,8400$	$0,1116 \cdot 10^{-7}$
1400	$0,1020 \cdot 10^{-5}$	$0,4501 \cdot 10^{-6}$	$2,2660$	$0,7603 \cdot 10^{-7}$
1500	$0,5067 \cdot 10^{-5}$	$0,1885 \cdot 10^{-5}$	$2,6990$	$0,4016 \cdot 10^{-6}$
1600	$0,2074 \cdot 10^{-4}$	$0,6615 \cdot 10^{-5}$	$3,1350$	$0,1726 \cdot 10^{-5}$
1700	$0,7131 \cdot 10^{-4}$	$0,2005 \cdot 10^{-4}$	$3,5550$	$0,6250 \cdot 10^{-5}$
1800	$0,2135 \cdot 10^{-3}$	$0,5383 \cdot 10^{-4}$	$3,9670$	$0,1964 \cdot 10^{-4}$
1900	$0,5667 \cdot 10^{-3}$	$0,1303 \cdot 10^{-3}$	$4,3630$	$0,5475 \cdot 10^{-4}$
2000	$0,1371 \cdot 10^{-2}$	$0,2892 \cdot 10^{-3}$	$4,7410$	$0,1378 \cdot 10^{-3}$
2100	$0,3035 \cdot 10^{-2}$	$0,5954 \cdot 10^{-3}$	$5,0970$	$0,3176 \cdot 10^{-3}$
2200	$0,6240 \cdot 10^{-2}$	$0,1149 \cdot 10^{-2}$	$5,4300$	$0,6797 \cdot 10^{-3}$
2300	$0,1203 \cdot 10^{-1}$	$0,2094 \cdot 10^{-2}$	$5,7460$	$0,1361 \cdot 10^{-2}$
2400	$0,2195 \cdot 10^{-1}$	$0,3634 \cdot 10^{-2}$	$6,0390$	$0,2573 \cdot 10^{-2}$
2500	$0,3810 \cdot 10^{-1}$	$0,6037 \cdot 10^{-2}$	$6,3110$	$0,4625 \cdot 10^{-2}$
2600	$0,6333 \cdot 10^{-1}$	$0,9649 \cdot 10^{-2}$	$6,5630$	$0,7947 \cdot 10^{-2}$
2700	$0,1013$	$0,1490 \cdot 10^{-1}$	$6,7940$	$0,1312 \cdot 10^{-1}$
2800	$0,1565$	$0,2233 \cdot 10^{-1}$	$7,0060$	$0,2091 \cdot 10^{-1}$
2900	$0,2345$	$0,3256 \cdot 10^{-1}$	$7,2020$	$0,3226 \cdot 10^{-1}$
3000	$0,3417$	$0,4628 \cdot 10^{-1}$	$7,3820$	$0,4841 \cdot 10^{-1}$

* G. B. Sinyarev and M. V. Dobrovol'skiy. Zhidkostnyye raketnyye dvigateli [Liquid-propellant Rocket Engines], Oborongiz [State Publishing House of the Defense Industry], 1957.

$K_{p4} = \dfrac{p_{NO}^2}{p_{N_2}p_{O_2}}$	$K_{p5} = \dfrac{p_H^3}{p_{H_2}}$	$K_{p6} = \dfrac{p_O^2}{p_{O_2}}$	$K_{p7} = \dfrac{p_N^2}{p_{N_2}}$	$T°\,K$
$0,6653 \cdot 10^{-30}$	$0,1813 \cdot 10^{-70}$	$0,8191 \cdot 10^{-80}$	$0,216 \cdot 10^{-118}$	300
$0,4898 \cdot 10^{-22}$	$0,1811 \cdot 10^{-51}$	$0,3084 \cdot 10^{-58}$	$0,3359 \cdot 10^{-87}$	400
$0,2587 \cdot 10^{-17}$	$0,4899 \cdot 10^{-40}$	$0,2944 \cdot 10^{-45}$	$0,1879 \cdot 10^{-68}$	500
$0,3648 \cdot 10^{-14}$	$0,2153 \cdot 10^{-32}$	$0,1387 \cdot 10^{-36}$	$0,6218 \cdot 10^{-56}$	600
$0,6489 \cdot 10^{-12}$	$0,6425 \cdot 10^{-27}$	$0,2240 \cdot 10^{-30}$	$0,5633 \cdot 10^{-47}$	700
$0,3163 \cdot 10^{-10}$	$0,8426 \cdot 10^{-23}$	$0,1034 \cdot 10^{-25}$	$0,3010 \cdot 10^{-40}$	800
$0,6495 \cdot 10^{-9}$	$0,1369 \cdot 10^{-19}$	$0,4450 \cdot 10^{-22}$	$0,5230 \cdot 10^{-35}$	900
$0,7302 \cdot 10^{-8}$	$0,5148 \cdot 10^{-17}$	$0,3631 \cdot 10^{-19}$	$0,8239 \cdot 10^{-31}$	1000
$0,5277 \cdot 10^{-7}$	$0,6676 \cdot 10^{-15}$	$0,8820 \cdot 10^{-17}$	$0,2262 \cdot 10^{-27}$	1100
$0,2752 \cdot 10^{-6}$	$0,3886 \cdot 10^{-13}$	$0,8630 \cdot 10^{-15}$	$0,1673 \cdot 10^{-24}$	1200
$0,1112 \cdot 10^{-5}$	$0,1220 \cdot 10^{-11}$	$0,4191 \cdot 10^{-13}$	$0,4503 \cdot 10^{-22}$	1300
$0,3680 \cdot 10^{-5}$	$0,2358 \cdot 10^{-10}$	$0,1173 \cdot 10^{-11}$	$0,5478 \cdot 10^{-20}$	1400
$0,1039 \cdot 10^{-4}$	$0,3087 \cdot 10^{-9}$	$0,2113 \cdot 10^{-10}$	$0,3527 \cdot 10^{-18}$	1500
$0,2675 \cdot 10^{-4}$	$0,2944 \cdot 10^{-8}$	$0,2657 \cdot 10^{-9}$	$0,1354 \cdot 10^{-16}$	1600
$0,5738 \cdot 10^{-4}$	$0,2162 \cdot 10^{-7}$	$0,2486 \cdot 10^{-8}$	$0,3393 \cdot 10^{-15}$	1700
$0,1170 \cdot 10^{-3}$	$0,1277 \cdot 10^{-6}$	$0,1819 \cdot 10^{-7}$	$0,5961 \cdot 10^{-14}$	1800
$0,2213 \cdot 10^{-3}$	$0,6267 \cdot 10^{-6}$	$0,1080 \cdot 10^{-6}$	$0,7761 \cdot 10^{-13}$	1900
$0,3926 \cdot 10^{-3}$	$0,2631 \cdot 10^{-5}$	$0,5376 \cdot 10^{-6}$	$0,7829 \cdot 10^{-12}$	2000
$0,6595 \cdot 10^{-3}$	$0,9658 \cdot 10^{-5}$	$0,2299 \cdot 10^{-5}$	$0,6349 \cdot 10^{-11}$	2100
$0,1057 \cdot 10^{-2}$	$0,3155 \cdot 10^{-4}$	$0,8624 \cdot 10^{-5}$	$0,4263 \cdot 10^{-10}$	2200
$0,1625 \cdot 10^{-2}$	$0,9313 \cdot 10^{-4}$	$0,2885 \cdot 10^{-4}$	$0,2429 \cdot 10^{-9}$	2300
$0,2410 \cdot 10^{-2}$	$0,2516 \cdot 10^{-3}$	$0,8738 \cdot 10^{-4}$	$0,1198 \cdot 10^{-8}$	2400
$0,3391 \cdot 10^{-2}$	$0,6284 \cdot 10^{-3}$	$0,2423 \cdot 10^{-3}$	$0,5206 \cdot 10^{-8}$	2500
$0,4840 \cdot 10^{-2}$	$0,1464 \cdot 10^{-2}$	$0,6215 \cdot 10^{-3}$	$0,2023 \cdot 10^{-7}$	2600
$0,6592 \cdot 10^{-2}$	$0,3207 \cdot 10^{-2}$	$0,1487 \cdot 10^{-2}$	$0,7114 \cdot 10^{-7}$	2700
$0,8786 \cdot 10^{-2}$	$0,6649 \cdot 10^{-2}$	$0,3345 \cdot 10^{-2}$	$0,2289 \cdot 10^{-6}$	2800
$0,1148 \cdot 10^{-1}$	$0,1312 \cdot 10^{-1}$	$0,7117 \cdot 10^{-2}$	$0,6797 \cdot 10^{-6}$	2900
$0,1472 \cdot 10^{-1}$	$0,2475 \cdot 10^{-1}$	$0,1441 \cdot 10^{-1}$	$0,1879 \cdot 10^{-5}$	3000

T° K	$K_{p1} = \dfrac{p_{CO} p_{O_2}^{1/2}}{p_{CO_2}}$	$K_{p2} = \dfrac{p_{H_2} p_{O_2}^{1/2}}{p_{H_2O}}$	$K_3 = \dfrac{p_{CO} p_{H_2O}}{p_{CO_2} p_{H_2}}$	$K_{p3} = \dfrac{p_{OH} p_{H_2}^{1/2}}{p_{H_2O}}$
3100	0,4854	$0,6436 \cdot 10^{-1}$	7,5430	$0,7074 \cdot 10^{-1}$
3200	0,6744	$0,8770 \cdot 10^{-1}$	7,6900	0,1009
3300	0,9179	0,1173	7,8210	0,1410
3400	1,2260	0,1544	7,9410	0,1933
3500	1,6100	0,2000	8,0480	0,2601
3600	2,0810	0,2556	8,1430	0,3444
3700	2,6520	0,3222	8,2280	0,4492
3800	3,3340	0,4017	8,2990	0,5780
3900	4,1410	0,4951	8,3640	0,7343
4000	5,0870	0,6042	8,4180	0,9217
4100	6,1810	0,7303	8,4650	1,1450
4200	7,4420	0,8750	8,5060	1,4070
4300	8,8740	1,0400	8,5310	1,7130
4400	10,5000	1,2280	8,5570	2,0670
4500	12,5300	1,4370	8,5760	2,4750
4600	14,3600	1,6730	8,5860	2,9400
4700	16,6200	1,9340	8,5920	3,4680
4800	19,1100	2,2240	8,5920	4,0610
4900	21,8400	2,5430	8,5880	4,7290
5000	24,8300	2,8940	8,5780	5,4730
5100	28,0600	3,2760	8,5670	6,2960
5200	31,5600	3,6940	8,5490	7,2090
5300	35,3700	4,1460	8,5310	8,2130
5400	39,4100	4,6340	8,5040	9,3110
5500	43,7500	5,1600	8,4780	10,5100
5600	48,3700	5,7270	8,4450	11,8100
5700	53,2700	6,3330	8,4120	13,2300
5800	58,4800	6,9810	8,3770	14,7600
5900	63,9600	7,6700	8,3390	16,4000
6000	69,7400	8,4050	8,2990	18,1600

$K_{p4} = \dfrac{p_{NO}^2}{p_{N_2} p_{O_2}}$	$K_{p5} = \dfrac{p_H^2}{p_{H_2}}$	$K_{p6} = \dfrac{p_O^2}{p_{O_2}}$	$K_{p7} = \dfrac{p_N^2}{p_{N_2}}$	$T° K$
$0,1858 \cdot 10^{-1}$	$0,4485 \cdot 10^{-1}$	$0,2786 \cdot 10^{-1}$	$0,4866 \cdot 10^{-5}$	3100
$0,2310 \cdot 10^{-1}$	$0,7836 \cdot 10^{-1}$	$0,5174 \cdot 10^{-1}$	$0,1189 \cdot 10^{-4}$	3200
$0,2833 \cdot 10^{-1}$	$0,1324$	$0,9253 \cdot 10^{-1}$	$0,2751 \cdot 10^{-4}$	3300
$0,3431 \cdot 10^{-1}$	$0,2170$	$0,1600$	$0,6064 \cdot 10^{-4}$	3400
$0,4115 \cdot 10^{-1}$	$0,3459$	$0,2680$	$0,1278 \cdot 10^{-3}$	3500
$0,4882 \cdot 10^{-1}$	$0,5374$	$0,4364$	$0,2587 \cdot 10^{-3}$	3600
$0,5736 \cdot 10^{-1}$	$0,8156$	$0,6926$	$0,5042 \cdot 10^{-3}$	3700
$0,6677 \cdot 10^{-1}$	$1,2120$	$1,0730$	$0,9491 \cdot 10^{-3}$	3800
$0,7720 \cdot 10^{-1}$	$1,7630$	$1,6240$	$0,1731 \cdot 10^{-2}$	3900
$0,8851 \cdot 10^{-1}$	$2,5190$	$2,4060$	$0,3063 \cdot 10^{-2}$	4000
$0,1006$	$3,5380$	$3,5050$	$0,5276 \cdot 10^{-2}$	4100
$0,1141$	$4,8890$	$5,0100$	$0,8857 \cdot 10^{-2}$	4200
$0,1283$	$6,6580$	$7,0460$	$0,1452 \cdot 10^{-1}$	4300
$0,1436$	$8,9390$	$9,7540$	$0,2330 \cdot 10^{-1}$	4400
$0,1598$	$11,8500$	$13,3100$	$0,3680 \cdot 10^{-1}$	4500
$0,1770$	$15,5200$	$17,9800$	$0,5641 \cdot 10^{-1}$	4600
$0,1952$	$20,0800$	$23,8300$	$0,8541 \cdot 10^{-1}$	4700
$0,2143$	$25,7200$	$31,3200$	$0,1272$	4800
$0,2343$	$32,6200$	$40,7000$	$0,1863$	4900
$0,2553$	$40,9900$	$52,3400$	$0,2688$	5000
$0,2771$	$51,0100$	$66,6500$	$0,3828$	5100
$0,2997$	$62,9800$	$84,1000$	$0,5379$	5200
$0,3234$	$77,1400$	$105,2000$	$0,7464$	5300
$0,3478$	$93,7800$	$130,5000$	$1,0230$	5400
$0,3731$	$113,2000$	$160,6000$	$1,3890$	5500
$0,3988$	$135,7000$	$196,3000$	$1,8640$	5600
$0,4258$	$161,7000$	$238,1000$	$2,4750$	5700
$0,4533$	$191,5000$	$287,0000$	$3,2610$	5800
$0,4812$	$225,5000$	$343,8000$	$4,2560$	5900
$0,5100$	$264,0000$	$409,4000$	$5,5080$	6000

APPENDIX III

COMMON LOGARITHMS FOR CONSTANTS OF CHEMICAL EQUILIBRIUM FOR PRODUCTS OF FUEL COMBUSTION*

T K	$\lg K_{p1} = \lg \frac{P_{CO}P_{O_2}^{1/2}}{P_{CO_2}}$	$\Delta \lg K_{p1}$	$\lg K_{p2} = \lg \frac{P_{H_2}P_{O_2}^{1/2}}{P_{H_2O}}$	$\Delta \lg K_{p2}$	$\lg K_{3a} = \lg \frac{P_{CO}P_{H_2O}}{P_{CO_2}P_{H_2}}$	$\Delta \lg K_{3a}$	$\lg K_{p3} = \frac{P_{O_2}P_{H_2O}^{1/2}}{P_{H_2O}}$	$\Delta \lg K_{p3}$
300	−44,7389	12,3295	−39,7860	10,5464	−4,9529	1,7831	−46,2890	12,3814
400	−32,4094	7,4044	−29,2396	6,3542	−3,1698	1,0502	−33,9076	7,4539
500	−25,0050	4,9407	−22,8854	4,2534	−2,1196	0,6873	−26,4537	4,9852
600	−20,0643	3,5268	−18,6320	3,0494	−1,4323	0,4774	−21,4685	3,5706
700	−16,5375	2,6434	−15,5826	2,2949	−0,9549	0,3485	−17,8979	2,6846
800	−13,8941	2,0541	−13,2877	1,7909	−0,6064	0,2632	−15,2133	2,0923
900	−11,8400	1,6415	−11,4968	1,4377	−0,3432	0,2038	−13,1210	1,6779
1000	−10,1985	1,3412	−10,0591	1,1779	−0,1394	0,1683	−11,4431	1,3735
1100	−8,8573	1,1160	−8,8812	0,9838	0,0239	0,1322	−10,0696	1,1463
1200	−7,7413	0,9428	−7,8974	0,8343	0,1561	0,1065	−8,9233	0,9710
1300	−6,7985	0,8009	−7,0631	0,7184	0,2646	0,0905	−7,9523	0,8333
1400	−5,9916	0,6980	−6,3467	0,6220	0,3551	0,0760	−7,1190	0,7228
1500	−5,2936	0,6194	−5,7247	0,5452	0,4311	0,0652	−6,3962	0,6331
1600	−4,6832	0,5363	−5,1795	0,4817	0,4963	0,0546	−5,7631	0,5590
1700	−4,1469	0,4763	−4,6978	0,4288	0,5509	0,0475	−5,2041	0,4973
1800	−3,6706	0,4255	−4,2690	0,3841	0,5984	0,0414	−4,7068	0,4452
1900	−3,2451	0,3821	−3,8849	0,3461	0,6398	0,0360	−4,2616	0,4009
2000	−2,8630	0,3451	−3,5388	0,3136	0,6758	0,0315	−3,8607	0,3629
2100	−2,5179	0,3131	−3,2252	0,2856	0,7073	0,0275	−3,4978	0,3301
2200	−2,2048	0,2852	−2,9396	0,2606	0,7348	0,0246	−3,1677	0,3016
2300	−1,9196	0,2610	−2,6790	0,2394	0,7594	0,0216	−2,8661	0,2766
2400	−1,6586	0,2395	−2,4396	0,2204	0,7810	0,0191	−2,5895	0,2546
2500	−1,4191	0,2207	−2,2192	0,2087	0,8001	0,0170	−2,3349	0,2351
2600	−1,1984	0,2039	−2,0155	0,1889	0,8171	0,0150	−2,0998	0,2178
2700	−0,9945	0,1900	−1,8266	0,1755	0,8321	0,0135	−1,8820	0,2024
2800	−0,8066	0,1756	−1,6511	0,1637	0,8456	0,0119	−1,6796	0,1885
2900	−0,6299	0,1635	−1,4874	0,1528	0,8575	0,0107	−1,4911	0,1760
3000	−0,4664	0,1525	−1,3346	0,1432	0,8682	0,0093	−1,3151	0,1648

* G. B. Sinyarev and M. V. Dobrovol'skiy. Zhidkostnyye raketnyye dvigateli [Liquid-propellant Rocket Engines], Oborongiz [State Publishing House of the Defense Industry], 1957.

$\lg K_{p4} = \lg \frac{p^2_{NO}}{p_N \cdot p_{O_2}}$	$\Delta \lg K_{p4}$	$\lg K_{p5} = \lg \frac{p^2_H}{p_{H_2}}$	$\Delta \lg K_{p5}$	$\lg K_{p6} = \lg \frac{p^2_O}{p_{O_2}}$	$\Delta \lg K_{p6}$	$\lg K_{p7} = \lg \frac{p^2_N}{p_{N_2}}$	$\Delta \lg K_{p7}$	T K
—30,1770	7,8670	—70,7414	18,9993	—80,0867	21,5758	—118,6656	31,1918	300
—22,3100	4,7228	—51,7421	11,4322	—58,5109	12,9798	—87,4738	17,7479	400
—17,5872	3,1492	—40,3099	7,6430	—45,5311	8,6731	—68,7259	12,5195	500
—14,4380	2,2502	—32,6669	5,4748	—38,8580	6,2081	—56,2064	8,9572	600
—12,1878	2,6880	—27,1921	4,1177	—30,6499	4,6645	—47,2492	6,7278	700
—10,4998	1,3124	—23,0744	3,2108	—25,9854	3,6339	—40,5214	5,2399	800
—9,1874	1,0506	—19,8636	2,5753	—22,3515	2,9115	—35,2815	4,1974	900
—8,1366	0,8590	—17,2883	2,1128	—19,4400	2,3855	—31,0841	3,4386	1000
—7,2776	0,7172	—15,1755	1,7650	—17,0545	1,9905	—27,6455	2,8689	1100
—6,5604	0,6064	—13,4105	1,4970	—15,0640	1,6863	—24,7766	2,4301	1200
—5,9540	0,5196	—11,9135	1,2960	—13,3777	1,4470	—22,3465	2,0851	1300
—5,4342	0,4508	—10,6275	1,1170	—11,9307	1,2555	—20,2614	1,8066	1400
—4,9836	0,3944	—9,5105	0,9794	—10,6752	1,0996	—18,4526	1,5842	1500
—4,5892	0,3480	—8,5311	0,8659	—9,5756	0,9712	—16,8684	1,3990	1600
—4,2412	0,3094	—7,6652	0,7711	—8,6044	0,8641	—15,4694	1,2447	1700
—3,9318	0,2766	—6,8941	0,6912	—7,7403	0,7738	—14,2247	1,1146	1800
—3,6550	0,2490	—6,2029	0,6231	—6,9665	0,6970	—13,1101	1,0038	1900
—3,4060	0,2252	—5,5798	0,5647	—6,2695	0,6311	—12,1063	0,9090	2000
—3,1808	0,2048	—5,0151	0,5141	—5,6384	0,5741	—11,1973	0,8270	2100
—2,9760	0,1868	—4,5010	0,4701	—5,0643	0,5245	—10,3703	0,7556	2200
—2,7892	0,1712	—4,0309	0,4315	—5,5398	0,4812	—9,6147	0,6929	2300
—2,6180	0,1574	—3,5994	0,3976	—4,0586	0,4429	—8,9218	0,6363	2400
—2,4606	0,1454	—3,2018	0,3674	—3,6157	0,4091	—8,2855	0,5895	2500
—2,3152	0,1342	—2,8344	0,3406	—3,2066	0,3789	—7,6940	0,5461	2600
—2,1810	0,1248	—2,4938	0,3166	—2,8277	0,3521	—7,1479	0,5075	2700
—2,0562	0,1156	—2,1772	0,2951	—2,4756	0,3279	—6,6404	0,4727	2800
—1,9404	0,1082	—1,8821	0,2757	—2,1477	0,3062	—6,1577	0,4416	2900
—1,8322	0,1012	—1,6064	0,2562	—1,8415	0,2865	—5,7261	0,4133	3000

$\lg K_{P4} = \dfrac{p_{NO}^2}{p_N \cdot p_{O_2}}$	$\Delta \lg K_{P4}$	$\lg K_{P5} = \lg \dfrac{p_H^2}{p_{H_2}}$	$\Delta \lg K_{P5}$	$\lg K_{P6} = \lg \dfrac{p_O^2}{p_{O_2}}$	$\Delta \lg K_{P6}$	$\lg K_{P7} = \lg \dfrac{p_N^2}{p_{N_2}}$	$\Delta \lg K_{P7}$	$T^\circ K$
—1,7310	0,0946	—1,3482	0,2423	—1,5550	0,2688	—5,3128	0,3878	3100
—1,6364	0,0888	—1,1059	2,2278	—1,2862	0,2525	—4,9250	0,3645	3200
—1,5476	0,0830	—0,8781	0,2146	—1,0337	0,2377	—4,5605	0,3433	3300
—1,4646	0,0790	—0,6635	0,2025	—0,7960	0,2242	—4,2172	0,3239	3400
—1,3856	0,0742	—0,4610	0,1913	—0,5718	0,2118	—3,8933	0,3061	3500
—1,3114	0,0700	—0,2697	0,1812	—0,3600	0,2005	—3,5872	0,2898	3600
—1,2414	0,0660	—0,0885	0,1717	—0,1595	0,1899	—3,2974	0,2747	3700
—1,1754	0,C630	0,0832	0,1630	0,0304	0,1802	—3,0227	0,2609	3800
—1,1124	0,0594	0,2462	0,1550	0,2106	0,1712	—2,7618	0,2480	3900
—1,0530	0,0564	0,4012	0,1475	0,3818	0,1629	—2,5138	0,2361	4000
—0,9666	0,0538	0,5487	0,1405	0,5447	0,1552	—2,2777	0,2250	4100
—0,9428	0,0512	0,6892	0,1341	0,6999	0,1480	—2,0527	0,2148	4200
—0,8916	0,0488	0,8233	0,1280	0,8479	0,1413	—1,8379	0,2052	4300
—0,8428	0,0464	0,9513	0,1223	0,9892	0,1351	—1,6327	0,1962	4400
—0,7964	0,0444	1,0736	0,1171	1,1243	0,1292	—1,4365	0,1879	4500
—0,7520	0,0424	1,1907	0,1122	1,2535	0,1237	—1,2486	0,1801	4600
—0,7096	0,0406	1,3029	0,1075	1,3772	0,1186	—1,0685	0,1728	4700
—0,6690	0,0388	1,4104	0,1031	1,4958	0,1138	—0,8957	0,1659	4800
—0,6302	0,0372	1,5135	0,0991	1,6096	0,1092	—0,7296	0,1595	4900
—0,5930	0,0356	1,6126	0,0951	1,7188	0,1050	—0,5703	0,1533	5000
—0,5574	0,0342	1,7077	0,0915	1,8238	0,1010	—0,4170	0,1477	5100
—0,5232	0,0330	1,7992	0,0881	1,9248	0,0972	—0,2698	0,1423	5200
—0,4902	0,0316	1,8873	0,0848	2,0220	0,0935	—0,1270	0,1372	5300
—0,4586	0,0304	1,9721	0,0818	2,1156	0,0902	0,0102	0,1324	5400
—0,4282	0,0290	2,0539	0,0788	2,2058	0,0870	0,1426	0,1278	5500
—0,3992	0,0284	2,1327	0,0760	2,2928	0,0640	0,2704	0,1235	5600
—0,3708	0,0272	2,2087	0,0735	2,3768	0,0811	0,3939	0,1195	5700
—0,3436	0,0260	2,2822	0,0709	2,4579	0,0784	0,5134	0,1156	5800
—0,3176	0,0252	2,3531	0,0685	2,5363	0,0758	0,6290	0,1120	5900
—0,2924		2,4216		2,6121		0,7410		6000

$T °K$	$\lg K_{p1}=\lg \dfrac{p_{CO}p_{O_2}^{1/2}}{p_{CO_2}}$	$\Delta \lg K_{p1}$	$\lg K_{p2}=\lg \dfrac{p_{H_2}p_{O_2}^{1/2}}{p_{H_2O}}$	$\Delta \lg K_{p2}$	$\lg K_{3a}=\lg \dfrac{p_{CO}p_{H_2O}}{p_{CO_2}p_{H_2}}$	$\Delta \lg K_{3a}$	$\lg K_{p3}=\dfrac{p_{OH}p_{H_2}^{1/2}}{p_{H_2O}}$	$\Delta \lg K_{p3}$
3100	—0,3139	0,1428	—1,1914	0,1344	0,8775	0,0084	—1,1503	0,1545
3200	—0,1711	0,1339	—1,0570	0,1265	0,8859	0,0074	—0,9958	0,1452
3300	—0,0372	0,1257	—0,9305	0,1191	0,8933	0,0069	—0,8506	0,1363
3400	0,0885	0,1183	—0,8114	0,1125	0,8999	0,0058	—0,7138	0,1290
3500	0,2068	0,1115	—0,6989	0,1064	0,9057	0,0051	—0,5848	0,1219
3600	0,3153	0,1052	—0,5925	0,1007	0,9108	0,0045	—0,4629	0,1154
3700	0,4235	0,0994	—0,4918	0,0957	0,9153	0,0037	—0,3475	0,1094
3800	0,5229	0,0942	—0,3961	0,0908	0,9190	0,0034	—0,2381	0,1040
3900	0,6171	0,0893	—0,3053	0,0865	0,9224	0,0028	—0,1341	0,0987
4000	0,7064	0,0847	—0,2188	0,0823	0,9252	0,0024	—0,0354	0,0940
4100	0,7911	0,0806	—0,1365	0,0785	0,9276	0,0021	0,0586	0,0896
4200	0,8717	0,0764	—0,0580	0,0751	0,9297	0,0013	0,1482	0,0855
4300	0,9481	0,0731	0,0171	0,0718	0,9310	0,0013	0,2337	0,0817
4400	1,0212	0,0696	0,0889	0,0686	0,9323	0,0010	0,3154	0,0781
4500	1,0908	0,0664	0,1575	0,0659	0,9333	0,0005	0,3935	0,0748
4600	1,1572	0,0634	0,2234	0,0631	0,9338	0,0003	0,4683	0,0716
4700	1,2206	0,0607	0,2865	0,0607	0,9341	0,0000	0,5399	0,0688
4800	1,2813	0,0580	0,3472	0,0582	0,9341	—0,0002	0,6087	0,0660
4900	1,3393	0,0556	0,4054	0,0561	0,9339	—0,0005	0,6747	0,0635
5000	1,3949	0,0533	0,4615	0,0539	0,9334	—0,0007	0,7382	0,0610
5100	1,4482	0,0512	0,5155	0,0521	0,9327	—0,0008	0,7992	0,0587
5200	1,4994	0,0492	0,5675	0,0501	0,9319	—0,0010	0,8579	0,0566
5300	1,5485	0,0470	0,6176	0,0484	0,9309	—0,0013	0,9145	0,0545
5400	1,5956	0,0454	0,6660	0,0467	0,9296	—0,0014	0,9690	0,0526
5500	1,6409	0,0435	0,7127	0,0452	0,9282	—0,0015	0,0216	0,0508
5600	1,6846	0,0420	0,7579	0,0437	0,9267	—0,0017	1,0724	0,0492
5700	1,7265	0,0405	0,8016	0,0423	0,9249	—0,0018	1,1216	0,0474
5800	1,7670	0,0389	0,8439	0,0409	0,9231	—0,0020	1,1690	0,0459
5900	1,8059	0,0376	0,8848	0,0397	0,9211	—0,0021	1,2149	0,0445
6000	1,8435		0,9245		0,9190		1,2594	

In the columns for the differences in the common logarithms of the equilibrium constants ($\Delta \log K_{r1}$, $\Delta \log K_{r3}$, etc.) is presented the difference in the values of the common logarithms of the equilibrium constants for the subsequent and current temperatures. For example, in the line $T = 3400°K$ we are given the difference in the value of $\log K_r$ at 3500 and $3400°K$.

Using the difference tables, we can determine the values of the equilibrium constants (by linear interpolation of the logarithms of the equilibrium constants) regardless of the intermediate temperature.

APPENDIX IV

ENTROPY S_1 OF PRODUCTS OF FUEL COMBUSTION IN kcal/kmole°C*

$T°$ K	CO_2	H_2O	CO	OH	NO	H_2
298,16	51,061	45,106	47,301	43,888	50,339	31,211
300	51,116	45,154	47,342	43,934	50,384	31,253
400	53,815	47,490	49,352	45,978	52,436	33,250
500	56,113	49,314	50,927	47,553	54,048	34,809
600	58,109	50,903	52,238	48,840	55,392	36,084
700	59,895	52,269	53,373	49,927	56,556	37,167
800	61,507	53,490	54,379	50,877	57,589	38,108
900	62,980	54,599	55,287	51,723	58,520	38,946
1000	64,3310	55,6180	56,1160	52,4910	59,3700	39,7040
1100	65,5822	56,5712	56,8779	53,1949	60,1500	40,3963
1200	66,7461	57,4654	57,5837	53,8470	60,8715	41,0365
1300	67,8334	58,3090	58,2413	54,4559	61,5425	41,6334
1400	68,8532	59,1084	58,8569	55,0278	62,1696	42,1938
1500	69,8132	59,8687	59,4353	55,5675	62,7580	42,7227
1600	70,7200	60,5939	59,9806	56,0788	63,3122	43,2243
1700	71,5792	61,2873	60,4964	56,5650	63,8358	43,7016
1800	72,3955	61,9515	60,9857	57,0285	64,3319	44,1571
1900	73,1727	62,5887	61,4510	57,4714	64,8034	44,5931
2000	73,9145	63,2010	61,8945	57,8956	65,2524	45,0112
2100	74,6238	63,7900	62,3181	58,3027	65,6811	45,4130
2200	75,3034	64,3574	62,7234	58,6939	66,0912	45,7998
2300	75,9557	64,9045	63,1121	59,0705	66,4841	46,1728
2400	76,5828	65,4328	63,4854	59,4337	66,8613	46,5329
2500	77,1865	65,9434	63,8444	59,7842	67,2240	46,8812
2600	77,7687	66,4374	64,1902	60,1230	67,5732	47,2183
2700	78,3307	66,9159	64,5238	60,4506	67,9100	47,5451
2800	78,8740	67,3796	64,8458	60,7684	68,2351	47,8622
2900	79,3997	67,8294	65,1572	61,0764	68,5494	48,1702
3000	79,9090	68,2661	65,4586	61,3753	68,8537	48,4696

* G. B. Sinyarev and M. V. Dobrovol'skiy. Zhidkostnyye raketnyye dvigateli [Liquid-propellant Rocket Engines], Oborongiz [State Publishing House of the Defense Industry], 1957.

O_2	N_2	H	O	N	$T°K$
49,011	45,767	27,3927	38,4689	36,6145	298,16
49,056	45,809	27,4232	38,5010	36,6450	300
51,098	47,818	28,8524	39,9915	38,0742	400
52,728	49,385	29,9610	41,1308	39,1828	500
54,105	50,685	30,8667	42,0540	40,0885	600
55,303	51,805	31,6326	42,8307	40,8544	700
56,368	52,797	32,2959	43,5011	41,5177	800
57,327	53,692	32,8811	44,0914	42,1029	900
58,1990	54,5090	33,4045	44,6183	42,6263	1000
58,9983	55,2601	33,8780	45,0945	43,0998	1100
59,7364	55,9565	34,3103	45,5288	43,5321	1200
60,4220	56,6060	34,7079	45,9281	43,9297	1300
61,0622	57,2143	35,0761	46,2975	44,2979	1400
61,6628	57,7863	35,4188	46,6413	44,6406	1500
62,2287	58,3261	35,7395	46,9628	44,9613	1600
62,7640	58,8371	36,0407	47,2646	45,2625	1700
63,2719	59,3221	36,3246	47,5492	45,5464	1800
63,7555	59,7836	36,5932	47,8184	45,8151	1900
64,2172	60,2237	36,8480	48,0737	46,0699	2000
64,6590	60,6443	37,0904	48,3166	46,3124	2100
65,0829	61,0471	37,3215	48,5481	46,5436	2200
65,4904	61,4333	37,5424	48,7695	46,7646	2300
65,8828	61,8044	37,7538	48,9814	46,9763	2400
66,2614	62,1614	37,9566	49,1848	47,1794	2500
66,6272	62,5054	38,1515	49,3803	47,3747	2600
66,9812	62,8373	38,3390	49,5686	47,5628	2700
67,3241	63,1579	38,5196	49,7501	47,7443	2800
67,6566	63,4679	38,6940	49,9254	47,9197	2900
67,9797	63,7680	38,8624	50,0950	48,0894	3000

$T°K$	CO_2	H_2O	CO	OH	NO	H_2
3100	80,4029	68,6904	65,7506	61,6658	69,1484	48,7609
3200	80,8822	69,1029	66,0338	61,9482	69,4342	49,0447
3300	81,3180	69,5042	66,3087	62,2230	69,7117	49,3213
3400	81,8006	69,8949	66,5757	62,4906	69,9813	49,5911
3500	82,2414	70,2754	66,8354	62,7515	70,2434	49,8545
3600	82,6705	70,6463	67,0880	63,0059	70,4985	50,1119
3700	83,0886	71,0080	67,3340	63,2542	70,7469	50,3635
3800	83,4963	71,3609	67,5738	63,4966	70,9891	50,6097
3900	83,8941	71,7054	67,8076	63,7336	71,2252	50,8506
4000	84,2826	72,0420	68,0357	63,9652	71,4556	51,0866
4100	84,6620	72,3710	68,2584	64,1917.	71,6806	51,3178
4200	85,0329	72,6926	68,4760	64,4135	71,9005	51,5445
4300	85,3957	73,0071	68,6887	64,6306	72,1154	51,7668
4400	85,7507	73,3150	68,8966	64,8434	72,3256	51,9850
4500	86,0982	73,6164	69,1001	65,0518	72,5312	52,1992
4600	86,4386	73,9117	69,2993	65,2563	72,7326	52,4096
4700	86,7721	74,2011	69,4944	65,4569	72,9299	52,6164
4800	87,0991	74,4847	69,6855	65,6537	73,1232	52,8196
4900	87,4198	74,7629	69,8728	65,8470	73,3128	53,0194
5000	87,7344	75,0359	70,0565	66,0368	73,4987	53,2159
5100	88,0433	75,3038	70,2367	66,2233	73,6812	53,4093
5200	88,3466	75,5669	70,4136	66,4065	73,8602	53,5997
5300	88,6445	75,8254	70,5872	66,5867	74,0361	53,7871
5400	88,9373	76,0794	70,7576	66,7639	74,2088	53,9717
5500	89,2251	76,3290	70,9251	66,9382	74,3786	54,1536
5600	89,5081	76,5745	71,0897	67,1097	74,5454	54,3328
5700	89,7865	76,8159	71,2514	67,2785	74,7095	54,5095
5800	90,0601	77,0535	71,4105	67,4447	74,8709	54,6837
5900	90,3301	77,2873	71,5670	67,6084	75,0297	54,8555
6000	90,5955	77,5175	71,7209	67,7697	75,1860	55,0250

O₂	N₂	H	O	N	T° K
68,2936	64,0588	39,0253	50,2592	48,2539	3100
68,5990	64,3409	39,1830	50,4183	48,4135	3200
68,8963	64,6148	39,3359	50,5728	48,5687	3300
69,1859	64,8809	39,4842	50,7229	48,7197	3400
69,4683	65,1397	39,6282	50,8689	48,8669	3500
69,7439	65,3915	39,7681	51,0111	49,0105	3600
70,0128	65,6368	39,9043	51,1496	49,1508	3700
70,2756	65,8758	40,0368	51,2846	49,2880	3800
70,5323	66,1089	40,1658	51,4165	49,4223	3900
70,7834	66,3364	40,2916	51,5452	49,5539	4000
71,0290	66,5585	40,4142	51,6711	49,6830	4100
71,2693	66,7754	40,5340	51,7942	49,8099	4200
71,5046	66,9875	40,6509	51,9147	49,9345	4300
71,7351	67,1949	40,7651	52,0326	50,0572	4400
71,9609	67,3979	40,8767	52,1482	50,1779	4500
72,1822	67,5965	40,9859	52,2615	50,2969	4600
72,3993	67,7911	41,0928	52,3727	50,4142	4700
72,6122	67,9818	41,1973	52,4817	50,5299	4800
72,8210	68,1687	41,2998	52,5888	50,6442	4900
73,0261	68,3520	41,4002	52,6939	50,7571	5000
73,2273	68,5318	41,4985	52,7972	50,8687	5100
73,4250	68,7082	41,5950	52,8988	50,9790	5200
73,6192	68,8815	41,6896	52,9986	51,0882	5300
73,8100	69,0516	41,7825	53,0968	51,1962	5400
73,9975	69,2188	41,8736	53,1933	51,3032	5500
74,1819	69,3831	41,9632	53,2884	51,4092	5600
74,3632	69,5446	42,0511	53,3820	51,5142	5700
74,5416	69,7034	42,1375	53,4742	51,6183	5800
74,7171	69,8596	42,2224	53,5649	51,7215	5900
74,8898	70,0134	42,3059	53,6544	51,8238	6000

COMMON LOGARITHMS FOR CONSTANTS OF CHEMICAL EQUILIBRIUM FOR PRODUCTS OF INCOMPLETE FUEL COMBUSTION

$T°K$	1 Метан CH₄ $lg K_p = lg \frac{p_{H_2}^4}{p_{CH_4}}$	$lg K_p = lg \frac{p_{CO}^2 p_{H_2}^3}{p_{CO_2} p_{CH_4}}$	2 Ацетилен C₂H₂ $lg K_p = lg \frac{p_{H}^4}{p_{C_2H_2}}$	$lg K_p = lg \frac{p_{CO}^4 p_{H_2}}{p_{CO_2}^2 p_{C_2H_2}}$	3 Этилен C₂H₄ $lg K_p = lg \frac{p_{H}^4}{p_{C_2H_4}}$	$lg K_p = lg \frac{p_{CO}^4 p_{H_2}^2}{p_{CO_2}^2 p_{C_2H_4}}$	4 Этан C₂H₆ $lg K_p = lg \frac{p_{H_2}^3}{p_{C_2H_6}}$	$lg K_p = lg \frac{p_{CO}^2 p_{H_2}^3}{p_{CO_2} p_{C_2H_6}}$	5 Бензол C₆H₆ $lg K_p = lg \frac{p_{H_2}^3}{p_{C_6H_6}}$	$lg K_p = lg \frac{p_{CO}^{12} p_{H_2}^3}{p_{CO_2}^6 p_{C_6H_6}}$	$T°K$
300	—	—29,6859	—	—5,2588	—	—29,8082	—	—47,3837	—	—102,3830	300
400	-5,5377	—18,8198	+26,5150	—0,0589	+9,6074	—16,9616	—1,9506	—28,5194	+19,0158	—60,6907	400
500	—3,4585	—12,2102	20,6101	+3,1040	8,3639	—9,1422	+0,4580	—17,0481	17,0437	—35,4746	500
600	—2,0339	—7,7620	16,6801	5,2221	7,5818	—3,8762	2,1233	—9,3346	15,8154	—18,5586	600
700	—0,9808	—4,5551	13,8806	6,7312	7,0454	—0,1040	3,3704	—3,7790	14,9836	—6,4646	700
800	—0,1738	—2,1367	11,7856	7,8596	6,6629	+2,7369	3,3123	+0,3863	14,3670	+2,5891	800
900	+0,4671	—0,2494	10,1646	8,7319	6,3788	4,9462	5,0680	3,6352	13,9098	9,6117	900
1000	0,9901	+1,2705	8,8642	9,4218	6,1513	6,7089	5,6843	6,2418	13,5264	15,1984	1000
1100	1,4320	2,5176	7,8107	9,9833	5,9712	8,1438	6,2003	8,3729	13,2512	19,7688	1100
1200	1,7992	3,5546	6,9277	10,4402	5,8254	9,3378	6,6191	10,1316	13,0109	23,5483	1200
1300	2,1058	4,4248	6,1905	10,8333	5,6990	10,3388	6,9855	11,6253	12,8000	26,7191	1300
1400	2,3616	5,1602	5,5523	11,1515	5,5969	11,1961	7,2862	12,8855	12,6334	29,4311	1400
1500	2,5713	5,7828	5,0114	11,4364	5,5023	11,9274	7,5409	13,9650	12,4628	31,7379	1500

1) Methane CH_4; 2) acetylene C_2H_2; 3) ethylene C_2H_4; 4) ethane C_2H_6; 5) benzene C_6H_6.

THERMODYNAMIC CHARACTERISTICS OF CERTAIN GASES

ГГаз $T°K$	CH$_4$	C$_2$H$_2$	C$_2$H$_4$	C$_3$H$_6$	C$_6$H$_6$
A 1. Энергосодержание (полная энтальпия) I, ккал/кмоль					
300	—17871	+54212	+12519	—20210	+19870
400	—16945	55346	13690	—18795	22200
500	—15919	56592	15075	—17062	25170
600	—14743	57931	16705	—15104	28710
700	—13419	59330	18483	—12816	32710
80)	—11978	60804	20439	—10339	37040
900	—10409	62308	22529	—7686	41680
1000	—8709	63886	24729	—4816	46570
1100	—6939	65497	27039	—1796	51680
1200	—5049	67139	29429	+1314	56990
1300	—3119	68861	31889	4614	62480
1400	—1169	70561	34469	7964	68030
1500	+801	72361	37049	11384	73640
B 2. Энтропия S, ккал/кмоль °C					
300	44,56	48,07	52,52	54,94	64,607
400	47,18	51,30	55,92	58,99	71,267
500	49,47	54,07	59,00	62,83	77,927
600	51,65	56,51	61,95	66,43	84,327
700	53,69	58,67	64,70	69,91	90,447
800	55,61	60,633	67,29	73,25	96,227
900	57,46	62,402	69,74	76,38	101,707
1000	59,22	64,076	72,06	79,38	106,927
1100	60,86	65,603	74,27	82,22	111,747
1200	62,47	67,050	76,34	84,95	116,357
1300	64,01	68,404	78,32	87,56	120,767
1400	65,49	69,682	80,92	90,08	124,847
1500	65,94	70,890	82,02	92,50	128,807
C 3. Молярные теплоемкости μCp, ккал/кмоль °C					
300	8,543	10,500	10,44	12,66	19,68
400	9,743	11,978	12,91	15,69	26,71
500	11,130	12,947	15,15	18,66	32,80
600	12,537	13,709	17,11	21,32	37,74
700	13,875	14,352	18,74	23,69	41,71
800	15,091	14,925	20,185	25,807	45,06
900	16,210	15,441	21,454	27,686	47,77
1000	17,196	15,909	22,556	29,279	50,14
1100	18,083	16,343	23,516	30,632	52,11
1200	18,877	16,734	24,364	31,840	53,81
1300	19,576	17,088	25,123	32,956	55,27
1400	20,179	17,406	25,763	33,968	56,54
1500	20,700	17,692	26,336	34,883	57,59

1) Gas; A) 1. Energy content (total enthalpy) I_1, kcal/kmole;B) 2. Entropy S_1, kcal/kmole °C;C) 3. Molar heat capacities μCp_1, kcal/kmole °C.

AUXILIARY THERMODYNAMIC-FUNCTION DATA AT DIFFERENT EXPONENTS FOR ISENTROPY OR POLYTROPY EQUATIONS

k	$\frac{1}{k-1}$	$\frac{k}{k-1}$	$\frac{k+1}{2}$	$\frac{2}{k+1}$	$\frac{2}{k-1}$	$\frac{k+1}{k-1}$	$\frac{k-1}{k+1}$	$\sqrt{\frac{k+1}{k-1}}$	$\left(\frac{k+1}{2}\right)^{\frac{1}{k-1}}$
1,10	10,000	11,000	1,050	0,952	20,000	21,000	0,0476	4,582	1,628
1,11	9,090	10,090	1,055	0,948	18,182	19,182	0,0526	4,380	1,626
1,12	8,333	9,333	1,060	0,943	16,666	17,666	0,0566	4,103	1,625
1,13	7,692	8,692	1,065	0,939	15,384	16,384	0,0611	4,046	1,623
1,14	7,143	8,143	1,070	0,934	14,286	15,286	0,0654	3,890	1,621
1,15	6,666	7,666	1,075	0,930	13,333	14,333	0,0697	3,786	1,615
1,16	6,250	7,250	1,080	0,926	12,500	13,500	0,0741	3,674	1,618
1,17	5,882	6,882	1,085	0,921	11,765	12,765	0,0778	3,573	1,616
1,18	5,555	6,555	1,090	0,917	11,111	12,111	0,0826	3,479	1,615
1,19	5,263	6,263	1,095	0,913	10,526	11,526	0,0867	3,395	1,612
1,20	5,000	6,000	1,100	0,909	10,000	10,000	0,0909	3,316	1,610
1,21	4,762	5,762	1,105	0,904	9,524	10,524	0,9500	3,244	1,609
1,22	4,545	5,545	1,110	0,901	9,091	10,091	0,0991	3,178	1,607
1,23	4,348	5,348	1,115	0,896	8,696	9,696	0,1031	3,113	1,605
1,24	4,166	5,166	1,120	0,893	8,333	9,333	0,1071	3,055	1,603
1,25	4,000	5,000	1,125	0,888	8,000	9,000	0,1111	3,000	1,602
1,26	3,846	4,846	1,130	0,885	7,692	7,692	0,1150	2,949	1,600
1,27	3,704	4,704	1,135	0,881	7,408	8,408	0,1193	2,900	1,599
1,28	3,572	4,572	1,140	0,877	7,143	8,143	0,1252	2,854	1,597
1,29	3,448	4,448	1,145	0,873	6,896	7,896	0,1266	2,699	1,596
1,30	3,333	4,333	1,150	0,869	6,666	7,666	0,1317	2,768	1,593
1,32	3,125	4,125	1,160	0,862	5,250	7,250	0,1379	2,692	1,590
1,34	2,941	3,941	1,170	0,856	5,882	6,882	0,1453	2,623	1,587
1,36	2,777	3,777	1,180	0,847	5,554	6,555	0,1525	2,560	1,584
1,38	2,692	3,692	1,190	0,840	5,263	6,263	0,1596	2,502	1,581
1,40	2,500	3,500	1,200	0,833	5,000	6,000	0,1666	2,449	1,575
1,50	2,000	3,000	1,250	0,800	4,000	5,000	0,2000	2,236	1,563

$\left(\dfrac{2}{k+1}\right)^{-\frac{1}{k-1}}$	$\left(\dfrac{k+1}{2}\right)^{\frac{k}{k-1}}$	$\left(\dfrac{2}{k+1}\right)^{\frac{k}{k-1}}$	$\left(\dfrac{2}{k+1}\right)^{\frac{2}{k-1}}$	$\left(\dfrac{2}{k+1}\right)^{\frac{k+1}{k-1}}$	$\sqrt{2g\dfrac{k}{k-1}}$	$\sqrt{gk\left(\dfrac{2}{k+1}\right)^{\frac{k+1}{k-1}}}$	$\sqrt{2k^2\left(\dfrac{2}{k-1}\right)^{\frac{k+1}{k-1}}}$
0,6139	1,712	0,5841	0,3767	0,3588	14,6801	1,978	0,6590
0,6146	1,718	0,5821	0,3775	0,3579	14,0680	1,984	0,6640
0,6153	1,724	0,5804	0,3784	0,3571	13,3381	1,988	0,6690
0,6160	1,730	0,5784	0,3792	0,3562	13,0591	1,993	0,6440
0,6168	1,736	0,5762	0,3800	0,3553	12,6125	1,998	0,6790
0,6175	1,744	0,5744	0,3809	0,3545	12,2637	2,000	0,6848
0,6182	1,749	0,5723	0,3818	0,3536	11,9261	2,009	0,6898
0,6188	1,754	0,5705	0,3826	0,3527	11,6190	2,014	0,6946
0,6195	1,761	0,5685	0,3835	0,3519	11,1175	2,018	0,6997
0,6202	1,767	0,5665	0,3844	0,3510	11,0851	2,024	0,7052
0,6209	1,771	0,5645	0,3852	0,3501	10,8570	2,030	0,7104
0,6216	1,779	0,5621	0,3860	0,3493	10,6323	2,034	0,7153
0,6223	1,785	0,5603	0,3869	0,3485	10,4302	2,040	0,7204
0,6230	1,790	0,5588	0,3878	0,3477	10,2435	2,045	0,7251
0,6237	1,796	0,5569	0,3886	0,3469	10,0672	2,050	0,7302
0,6243	1,801	0,5549	0,3894	0,3462	9,9044	2,055	0,7356
0,6250	1,818	0,5532	0,3903	0,3454	9,7484	2,061	0,7404
0,6257	1,813	0,5513	0,3911	0,3446	9,6067	2,066	0,7453
0,6264	1,820	0,5494	0,3919	0,3437	9,4740	2,072	0,7500
0,6271	1,825	0,5475	0,3927	0,3430	9,3418	2,078	0,7552
0,6276	1,832	0,5457	0,3936	0,3423	9,2495	2,083	0,7598
0,6289	1,838	0,5421	0,3953	0,3407	8,9961	2,094	0,7695
0,6302	1,856	0,5396	0,3968	0,3392	8,7931	2,105	0,7792
0,6314	1,868	0,5352	0,3985	0,3377	8,6280	2,111	0,7906
0,6327	1,880	0,5317	0,1001	0,3362	8,4403	2,129	0,7992
0,6339	1,893	0,5283	0,4018	0,3348	8,2867	2,140	0,8102
0,6401	1,953	0,5120	0,4096	0,3277	7,6720	2,210	0,8586

Made in the USA
Columbia, SC
04 January 2022